Teacher, Student, and Parent
One-Stop Internet Resources

Log on to
booko.msscience.com

ONLINE STUDY TOOLS

- Section Self-Check Quizzes
- Interactive Tutor
- Chapter Review Tests
- Standardized Test Practice
- Vocabulary PuzzleMaker

ONLINE RESEARCH

- WebQuest Projects
- Prescreened Web Links
- Career Links
- Internet Labs

INTERACTIVE ONLINE STUDENT EDITION

- Complete Interactive Student Edition available at mhln.com

FOR TEACHERS

- Teacher Bulletin Board
- Teaching Today—Professional Development

SAFETY SYMBOLS

	HAZARD	EXAMPLES	PRECAUTION	REMEDY
DISPOSAL	Special disposal procedures need to be followed.	certain chemicals, living organisms	Do not dispose of these materials in the sink or trash can.	Dispose of wastes as directed by your teacher.
BIOLOGICAL	Organisms or other biological materials that might be harmful to humans	bacteria, fungi, blood, unpreserved tissues, plant materials	Avoid skin contact with these materials. Wear mask or gloves.	Notify your teacher if you suspect contact with material. Wash hands thoroughly.
EXTREME TEMPERATURE	Objects that can burn skin by being too cold or too hot	boiling liquids, hot plates, dry ice, liquid nitrogen	Use proper protection when handling.	Go to your teacher for first aid.
SHARP OBJECT	Use of tools or glassware that can easily puncture or slice skin	razor blades, pins, scalpels, pointed tools, dissecting probes, broken glass	Practice common-sense behavior and follow guidelines for use of the tool.	Go to your teacher for first aid.
FUME	Possible danger to respiratory tract from fumes	ammonia, acetone, nail polish remover, heated sulfur, moth balls	Make sure there is good ventilation. Never smell fumes directly. Wear a mask.	Leave foul area and notify your teacher immediately.
ELECTRICAL	Possible danger from electrical shock or burn	improper grounding, liquid spills, short circuits, exposed wires	Double-check setup with teacher. Check condition of wires and apparatus.	Do not attempt to fix electrical problems. Notify your teacher immediately.
IRRITANT	Substances that can irritate the skin or mucous membranes of the respiratory tract	pollen, moth balls, steel wool, fiberglass, potassium permanganate	Wear dust mask and gloves. Practice extra care when handling these materials.	Go to your teacher for first aid.
CHEMICAL	Chemicals can react with and destroy tissue and other materials	bleaches such as hydrogen peroxide; acids such as sulfuric acid, hydrochloric acid; bases such as ammonia, sodium hydroxide	Wear goggles, gloves, and an apron.	Immediately flush the affected area with water and notify your teacher.
TOXIC	Substance may be poisonous if touched, inhaled, or swallowed.	mercury, many metal compounds, iodine, poinsettia plant parts	Follow your teacher's instructions.	Always wash hands thoroughly after use. Go to your teacher for first aid.
FLAMMABLE	Flammable chemicals may be ignited by open flame, spark, or exposed heat.	alcohol, kerosene, potassium permanganate	Avoid open flames and heat when using flammable chemicals.	Notify your teacher immediately. Use fire safety equipment if applicable.
OPEN FLAME	Open flame in use, may cause fire.	hair, clothing, paper, synthetic materials	Tie back hair and loose clothing. Follow teacher's instruction on lighting and extinguishing flames.	Notify your teacher immediately. Use fire safety equipment if applicable.

 Eye Safety
Proper eye protection should be worn at all times by anyone performing or observing science activities.

 Clothing Protection
This symbol appears when substances could stain or burn clothing.

 Animal Safety
This symbol appears when safety of animals and students must be ensured.

 Handwashing
After the lab, wash hands with soap and water before removing goggles.

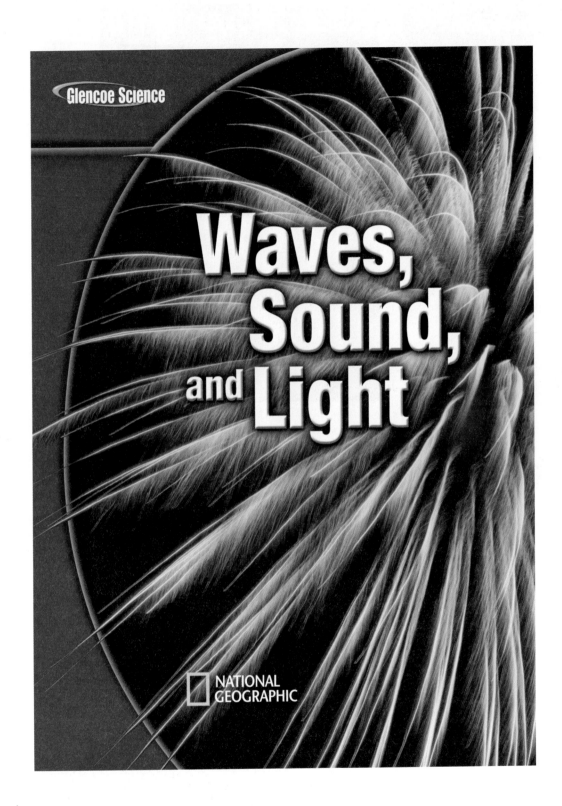

Glencoe Science

Waves, Sound, and Light

NATIONAL GEOGRAPHIC

Glencoe

New York, New York   Columbus, Ohio   Chicago, Illinois   Woodland Hills, California

Waves, Sound, and Light

The amount of light energy emitted determines the color of fireworks. Common substances used are strontium or lithium salts for red, calcium salts for orange, sodium compounds for yellow, barium chloride for green, copper chloride for blue, and strontium and copper compounds for purple.

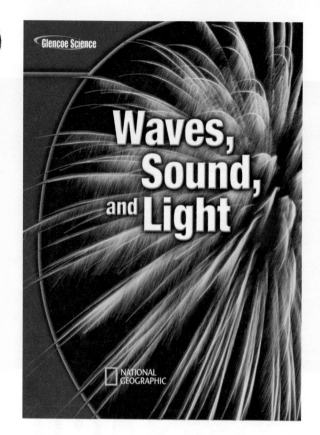

 **Glencoe**

The McGraw·Hill Companies

Send all inquiries to:
Glencoe/McGraw-Hill
8787 Orion Place
Columbus, OH 43240-4027

ISBN: 978-0-07-877840-7
MHID: 0-07-877840-9

Printed in the United States of America.

3 4 5 6 7 8 9 10 DOW 09

Authors

NATIONAL GEOGRAPHIC
Education Division
Washington, D.C.

Nicholas Hainen
Chemistry/Physics Teacher, Retired
Worthington City Schools
Worthington, OH

Dinah Zike
Educational Consultant
Dinah-Might Activities, Inc.
San Antonio, TX

Cathy Ezrailson
Science Department Head
Academy for Science and Health
Professions
Conroe, TX

Deborah Lillie
Math and Science Writer
Sudbury, MA

Series Consultants

CONTENT

Jack Cooper
Ennis High School
Ennis, TX

Carl Zorn, PhD
Staff Scientist
Jefferson Laboratory
Newport News, VA

MATH

Michael Hopper, DEng
Manager of Aircraft Certification
L-3 Communications
Greenville, TX

READING

Rachel Swaters-Kissinger
Science Teacher
John Boise Middle School
Warsaw, MO

SAFETY

Aileen Duc, PhD
Science 8 Teacher
Hendrick Middle School, Plano ISD
Plano, TX

Sandra West, PhD
Department of Biology
Texas State University-San Marcos
San Marcos, TX

ACTIVITY TESTERS

Nerma Coats Henderson
Pickerington Lakeview Jr. High
School
Pickerington, OH

Mary Helen Mariscal-Cholka
William D. Slider Middle School
El Paso, TX

**Science Kit and Boreal
Laboratories**
Tonawanda, NY

Series Reviewers

Desiree Bishop
Environmental Studies Center
Mobile County Public Schools
Mobile, AL

Tom Bright
Concord High School
Charlotte, NC

Anthony J. DiSipio, Jr.
8th Grade Science
Octorana Middle School
Atglen, PA

George Gabb
Great Bridge Middle School
Chesapeake Public Schools
Chesapeake, VA

Annette Parrott
Lakeside High School
Atlanta, GA

Karen Watkins
Perry Meridian Middle School
Indianapolis, IN

Clabe Webb
Permian High School
Ector County ISD
Odessa, TX

Kate Ziegler
Durant Road Middle School
Raleigh, NC

HOW TO...
Use Your Science Book

Why do I need my science book?

Why do I need my science book?

Have you ever been in class and not understood all of what was presented? Or, you understood everything in class, but at home, got stuck on how to answer a question? Maybe you just wondered when you were ever going to use this stuff?

These next few pages are designed to help you understand everything your science book can be used for . . . besides a paperweight!

Before You Read

- **Chapter Opener** Science is occurring all around you, and the opening photo of each chapter will preview the science you will be learning about. The **Chapter Preview** will give you an idea of what you will be learning about, and you can try the **Launch Lab** to help get your brain headed in the right direction. The **Foldables** exercise is a fun way to keep you organized.

- **Section Opener** Chapters are divided into two to four sections. The **As You Read** in the margin of the first page of each section will let you know what is most important in the section. It is divided into four parts. **What You'll Learn** will tell you the major topics you will be covering. **Why It's Important** will remind you why you are studying this in the first place! The **Review Vocabulary** word is a word you already know, either from your science studies or your prior knowledge. The **New Vocabulary** words are words that you need to learn to understand this section. These words will be in **boldfaced** print and highlighted in the section. Make a note to yourself to recognize these words as you are reading the section.

Glencoe Science

Waves, Sound, and Light

NATIONAL GEOGRAPHIC

As You Read

- **Headings** Each section has a title in large red letters, and is further divided into blue titles and small red titles at the beginnings of some paragraphs. To help you study, make an outline of the headings and subheadings.

- **Margins** In the margins of your text, you will find many helpful resources. The **Science Online** exercises and **Integrate** activities help you explore the topics you are studying. **MiniLabs** reinforce the science concepts you have learned.

- **Building Skills** You also will find an **Applying Math** or **Applying Science** activity in each chapter. This gives you extra practice using your new knowledge, and helps prepare you for standardized tests.

- **Student Resources** At the end of the book you will find **Student Resources** to help you throughout your studies. These include **Science, Technology,** and **Math Skill Handbooks,** an **English/Spanish Glossary,** and an **Index.** Also, use your **Foldables** as a resource. It will help you organize information, and review before a test.

- **In Class** Remember, you can always ask your teacher to explain anything you don't understand.

FOLDABLES ™
Study Organizer

Science Vocabulary Make the following Foldable to help you understand the vocabulary terms in this chapter.

STEP 1 Fold a vertical sheet of notebook paper from side to side.

STEP 2 Cut along every third line of only the top layer to form tabs.

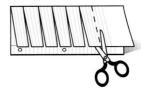

STEP 3 Label each tab with a vocabulary word from the chapter.

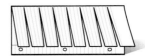

Build Vocabulary As you read the chapter, list the vocabulary words on the tabs. As you learn the definitions, write them under the tab for each vocabulary word.

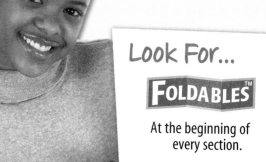

Look For...

FOLDABLES ™

At the beginning of every section.

In Lab

Working in the laboratory is one of the best ways to understand the concepts you are studying. Your book will be your guide through your laboratory experiences, and help you begin to think like a scientist. In it, you not only will find the steps necessary to follow the investigations, but you also will find helpful tips to make the most of your time.

- Each lab provides you with a **Real-World Question** to remind you that science is something you use every day, not just in class. This may lead to many more questions about how things happen in your world.

- Remember, experiments do not always produce the result you expect. Scientists have made many discoveries based on investigations with unexpected results. You can try the experiment again to make sure your results were accurate, or perhaps form a new hypothesis to test.

- Keeping a **Science Journal** is how scientists keep accurate records of observations and data. In your journal, you also can write any questions that may arise during your investigation. This is a great method of reminding yourself to find the answers later.

Look For...
- **Launch Labs** start every chapter.
- **MiniLabs** in the margin of each chapter.
- **Two Full-Period Labs** in every chapter.
- **EXTRA Try at Home Labs** at the end of your book.
- the **Web site** with laboratory demonstrations.

Before a Test

Admit it! You don't like to take tests! However, there *are* ways to review that make them less painful. Your book will help you be more successful taking tests if you use the resources provided to you.

- Review all of the **New Vocabulary** words and be sure you understand their definitions.

- Review the notes you've taken on your **Foldables,** in class, and in lab. Write down any question that you still need answered.

- Review the **Summaries** and **Self Check questions** at the end of each section.

- Study the concepts presented in the chapter by reading the **Study Guide** and answering the questions in the **Chapter Review.**

Look For...

- **Reading Checks** and **caption questions** throughout the text.
- the **Summaries** and **Self Check questions** at the end of each section.
- the **Study Guide** and **Review** at the end of each chapter.
- the **Standardized Test Practice** after each chapter.

Let's Get Started

To help you find the information you need quickly, use the Scavenger Hunt below to learn where things are located in Chapter 1.

1. What is the title of this chapter?

2. What will you learn in Section 1?

3. Sometimes you may ask, "Why am I learning this?" State a reason why the concepts from Section 2 are important.

4. What is the main topic presented in Section 2?

5. How many reading checks are in Section 1?

6. What is the Web address where you can find extra information?

7. What is the main heading above the sixth paragraph in Section 2?

8. There is an integration with another subject mentioned in one of the margins of the chapter. What subject is it?

9. List the new vocabulary words presented in Section 2.

10. List the safety symbols presented in the first Lab.

11. Where would you find a Self Check to be sure you understand the section?

12. Suppose you're doing the Self Check and you have a question about concept mapping. Where could you find help?

13. On what pages are the Chapter Study Guide and Chapter Review?

14. Look in the Table of Contents to find out on which page Section 2 of the chapter begins.

15. You complete the Chapter Review to study for your chapter test. Where could you find another quiz for more practice?

Teacher Advisory Board

The Teacher Advisory Board gave the editorial staff and design team feedback on the content and design of the Student Edition. They provided valuable input in the development of the 2008 edition of *Glencoe Science.*

John Gonzales
Challenger Middle School
Tucson, AZ

Rachel Shively
Aptakisic Jr. High School
Buffalo Grove, IL

Roger Pratt
Manistique High School
Manistique, MI

Kirtina Hile
Northmor Jr. High/High School
Galion, OH

Marie Renner
Diley Middle School
Pickerington, OH

Nelson Farrier
Hamlin Middle School
Springfield, OR

Jeff Remington
Palmyra Middle School
Palmyra, PA

Erin Peters
Williamsburg Middle School
Arlington, VA

Rubidel Peoples
Meacham Middle School
Fort Worth, TX

Kristi Ramsey
Navasota Jr. High School
Navasota, TX

Student Advisory Board

The Student Advisory Board gave the editorial staff and design team feedback on the design of the Student Edition. We thank these students for their hard work and creative suggestions in making the 2008 edition of *Glencoe Science* student friendly.

Jack Andrews
Reynoldsburg Jr. High School
Reynoldsburg, OH

Peter Arnold
Hastings Middle School
Upper Arlington, OH

Emily Barbe
Perry Middle School
Worthington, OH

Kirsty Bateman
Hilliard Heritage Middle School
Hilliard, OH

Andre Brown
Spanish Emersion Academy
Columbus, OH

Chris Dundon
Heritage Middle School
Westerville, OH

Ryan Manafee
Monroe Middle School
Columbus, OH

Addison Owen
Davis Middle School
Dublin, OH

Teriana Patrick
Eastmoor Middle School
Columbus, OH

Ashley Ruz
Karrer Middle School
Dublin, OH

The Glencoe middle school science Student Advisory Board taking a timeout at COSI, a science museum in Columbus, Ohio.

In each chapter, look for these opportunities for review and assessment:
- **Reading Checks**
- **Caption Questions**
- **Section Review**
- **Chapter Study Guide**
- **Chapter Review**
- **Standardized Test Practice**
- **Online practice at booko.msscience.com**

Get Ready to Read Strategies
- **Preview** 8A
- **Identify the Main Idea** 36A
- **New Vocabulary** 66A
- **Monitor** 96A

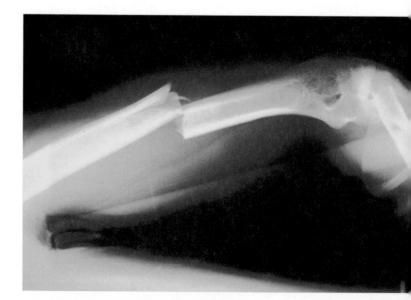

chapter 4 — Light, Mirrors, and Lenses—94

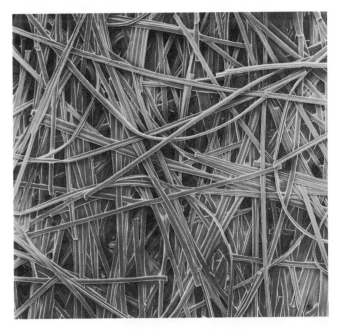

Student Resources

Cross-Curricular Readings/Labs

DVD available as a video lab

NATIONAL GEOGRAPHIC VISUALIZING

TIME SCIENCE AND Society

TIME SCIENCE AND HISTORY

Oops! Accidents in SCIENCE

SCIENCE Stats

Launch LAB

Mini LAB

Mini LAB Try at Home

Labs/Activities

One-Page Labs

Two-Page Labs

Design Your Own Labs

Applying Math

Applying Science

INTEGRATE

Science Online

Standardized Test Practice

Content Details

Let There Be Light

Figure 1 Thomas Edison conducted thousands of experiments to find the proper filament material for one of his greatest inventions—the electric lightbulb.

T here's a well-known expression that advises "if at first you don't succeed, try, try again." Fortunately, Thomas Edison lived by these words as he worked in his research laboratories developing over 1,000 patented inventions. Among the many items that his team developed are the phonograph, the first commercial electric light and power system, a motion picture camera, and the incandescent lamp. Edison's search for a suitable filament for the incandescent lamp demonstrates how he used the experimental method to guide his scientific research.

The Search for Filament Material

When electric current is passed through the filament or wire inside the lightbulb, the filament heats up and begins to glow. The problem for Edison and his team of researchers was finding a filament substance that would glow for a long time without incinerating (turning to ashes), fusing, or melting.

Before experimenting with filaments, Edison knew that he had to find a way to keep the materials in lightbulbs from incinerating. Oxygen is required for a substance to burn, so he removed the air from his lightbulb, creating a vacuum, around the filament. Then the search for the proper filament began.

Figure 2 Many of Edison's greatest inventions, including the phonograph and the electric lightbulb, were developed in his laboratory in Menlo Park, New Jersey. In fact, Edison was called "The Wizard of Menlo Park."

Experimentation and Improvement

Edison unsuccessfully experimented with more than 1,600 materials, including plant fibers, fishing line, hair, and platinum. Then, Edison and his team experimented with carbon, a nonmetallic element that was inexpensive and glowed when current was passed through it. Because carbon can't be shaped into a wire, Edison had to coat other substances with carbon to make the lightbulb filament. In 1879, one of Edison's researchers tested a thin piece of carbonized cotton. The tiny filament glowed for at least 13 hours before Edison increased the voltage and it burned out. The experiments carried out by Edison finally resulted in a useable lightbulb which Edison patented in 1880.

Lewis Latimer, an African American inventor, also used experimentation to make significant improvements to the lightbulb. He developed and patented a method for connecting the electrical wires and the carbon filament together in the base of the bulb in 1881 and a process to make a long-lasting carbon filament in 1882. Experimentation and improvements to electrical lighting continue today and longer-lasting lightbulbs are the result.

Figure 3 Edison designed an airless glass bulb in which to test filament materials.

Figure 4 Lewis Latimer significantly improved the carbon filament, making electric lightbulbs more efficient and durable.

Figure 5 Because of continued experimentation and improvements, modern incandescent lightbulbs, like those that help light this city, typically last for about 1,000 hours. Some specially designed bulbs last as long as 20,000 hours.

The Study of Matter and Energy

Edison and Latimer, like all scientists, attempted to answer questions by performing tests and recording the results. When you answer a question or solve a problem by conducting a test, you are taking the scientific approach.

Experiments with electricity and light are part of physical science, the study of matter and energy. Two of the main branches of physical science are chemistry and physics. Chemistry is the study of what substances are made of and how they change. Physics is the study of matter and energy, including light and sound.

Experimentation

Experiments must be carefully planned in order to insure the accuracy of the results. Scientists begin by defining what they expect the experiment to prove. Edison's filament experiments were designed to find which material would act as the best filament for an incandescent lightbulb. Edison tested filament materials by placing them in airless bulbs and then running electric current through them.

Variables and Controls in an Experiment

When scientists conduct experiments, they must make sure that only one factor affects the results of the experiment. The factor being changed is called the independent variable. The dependent variable is what is measured or observed to obtain the results of the experiment. In Edison's filament experiment, the independent variables were the different materials that were tested as filaments. The dependent variable was how long each of the tested substances glowed when electric current flowed through them.

The conditions that stay the same in an experiment are called constants. The constants in Edison's filament experiments

Steps for Experiments

1. Limit independent variables.
Only one independent variable should be used in any experiment.

2. Use a control.
There must be a sample group that is treated like the others except the independent variable isn't applied.

3. Repeat the experiment.
To insure that the results are valid, experiments must be repeated several times.

Figure 6 In this illustration, Edison (third from left) tests the electric light as his fellow researchers observe the results.

included the voltage applied and using the same type of bulb to surround each filament.

Edison changed a factor that should have been a constant, however, when he increased the voltage running through the carbonized cotton thread. Well-planned experiments also need a control—a sample that is treated like all the others except the independent variable isn't changed.

Interpreting Data

The observations and measurements that a scientist makes in an experiment are called data. Data must be carefully studied before questions can be answered or problems can be solved. Scientists repeat their experiments many times to make sure that their results are accurate.

Drawing Conclusions, Eliminating Biases

A conclusion is a statement that summarizes the results of the data that is obtained by the experiment. It is important that scientists are not influenced or biased by what they think the results will be or by what they want the results to be. A bias is a prejudice or an opinion. To avoid a biased conclusion it is important that scientists look at their data carefully and make sure their conclusion is based on their data. If more than one conclusion is possible, scientists often will conduct more tests to eliminate some of the possibilities or to find the best solution. Edison found several materials that glowed when a voltage was applied, but they were not suitable for lighting for various reasons. He found that carbon glowed when a voltage was applied and it had other qualities that made it a good choice for the filament. However, since carbon was brittle and did not form a wire, he had to keep experimenting to find the best material to support the carbon to make the filament.

"Results? Why, man, I have gotten lots of results! If I find 10,000 ways something won't work, I haven't failed. I am not discouraged, because every wrong attempt discarded is another step forward. Just because something doesn't do what you planned it to do doesn't mean it's useless.... Reverses should prove an incentive to great accomplishment.... There are no rules here, we're just trying to accomplish something."
-Thomas Edison

Figure 7 This quote from Thomas Edison is an example of a conclusion.

You Do It

Thomas Edison is only one of many inventors who conducted numerous experiments before creating a successful invention. Research the experiments that went into the invention of the telephone. How long did it take? How is the technology of the telephone that was used in 1900 different from the phone many people use today?

Waves

The BIG Idea

Waves transfer energy from place to place without transferring matter.

SECTION 1
What are waves?
Main Idea Waves can carry energy through matter or through empty space.

SECTION 2
Wave Properties
Main Idea Waves can be described by their amplitude, wavelength, and frequency.

SECTION 3
Wave Behavior
Main Idea Waves can change direction at the boundary between different materials.

Catch A Wave

On a breezy day in Maui, Hawaii, windsurfers ride the ocean waves. Waves carry energy. You can see the ocean waves in this picture, but there are other waves you cannot see, such as microwaves, radio waves, and sound waves.

Science Journal Write a paragraph about some places where you have seen water waves.

Start-Up Activities

Waves and Energy

It's a beautiful autumn day. You are sitting by a pond in a park. Music from a school marching band is carried to your ears by waves. A fish jumps, making waves that spread past a leaf that fell from a tree, causing the leaf to move. In the following lab, you'll observe how waves carry energy that can cause objects to move.

1. Add water to a large, clear, plastic plate to a depth of about 1 cm.
2. Use a dropper to release a single drop of water onto the water's surface. Repeat.
3. Float a cork or straw on the water.
4. When the water is still, repeat step 2 from a height of 10 cm, then again from 20 cm.
5. **Think Critically** In your Science Journal, record your observations. How did the motion of the cork depend on the height of the dropper?

Preview this chapter's content and activities at
booko.msscience.com

 **Waves** Make the following Foldable to compare and contrast the characteristics of transverse and compressional waves.

 **Fold** one sheet of paper lengthwise.

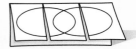 **Fold** into thirds.

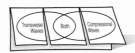 **Unfold and draw** overlapping ovals. **Cut** the top sheet along the folds.

STEP 4 **Label** the ovals as shown.

Construct a Venn Diagram As you read the chapter, list the characteristics unique to transverse waves under the left tab, those unique to compressional waves under the right tab, and those characteristics common to both under the middle tab.

Get Ready to Read

New Vocabulary

① Learn It! What should you do if you find a word you don't know or understand? Here are some suggested strategies:

1. Use context clues (from the sentence or the paragraph) to help you define it.
2. Look for prefixes, suffixes, or root words that you already know.
3. Write it down and ask for help with the meaning.
4. Guess at its meaning.
5. Look it up in the glossary or a dictionary.

② Practice It! Look at the word *rarefaction* in the following passage. See how context clues can help you understand its meaning.

Context Clue
Rarefactions are groups of molecules.

Context Clue
Rarefactions move away from a vibrating object.

Context Clue
Rarefactions are parts of a sound wave.

When the drumhead moves downward, the molecules near it have more room and can spread farther apart. This group of molecules that are farther apart is a rarefaction. The rarefaction also moves away from the drumhead. As the drumhead vibrates up and down, it forms a series of compressions and rarefactions that move away and spread out in all directions this series of compressions and rarefactions is a sound wave.

— *from page 11*

③ Apply It! Make a vocabulary bookmark with a strip of paper. As you read, keep track of words you do not know or want to learn more about.

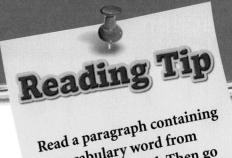

Reading Tip

Read a paragraph containing a vocabulary word from beginning to end. Then go back to determine the meaning of the word.

Target Your Reading

Use this to focus on the main ideas as you read the chapter.

1 **Before you read** the chapter, respond to the statements below on your worksheet or on a numbered sheet of paper.

- Write an **A** if you **agree** with the statement.
- Write a **D** if you **disagree** with the statement.

2 **After you read** the chapter, look back to this page to see if you've changed your mind about any of the statements.

- If any of your answers changed, explain why.
- Change any false statements into true statements.
- Use your revised statements as a study guide.

Science Online

Print out a worksheet of this page at booko.msscience.com

Before You Read A or D		Statement	After You Read A or D
	1	There can be no sound in outer space.	
	2	Light waves are mechanical waves that can travel only through matter.	
	3	Waves are produced by something that vibrates.	
	4	In air, sound waves travel faster than light waves.	
	5	The pitch of a sound wave depends on the frequency of the wave.	
	6	The only difference between microwaves and light waves is that microwaves have longer wavelengths.	
	7	A wave moves in a straight line, even if the speed of the wave changes.	
	8	Higher-pitched sounds bend more as they pass through an open door than lower-pitched sounds do.	
	9	When two waves overlap, they always cancel each other.	

What are waves?

What **You'll Learn**

- **Explain** the relationship among waves, energy, and matter.
- **Describe** the difference between transverse waves and compressional waves.

Why **It's Important**

Waves enable you to see and hear the world around you.

Review Vocabulary

energy: the ability to cause change

New Vocabulary

- wave
- mechanical wave
- transverse wave
- compressional wave
- electromagnetic wave

What is a wave?

When you are relaxing on an air mattress in a pool and someone does a cannonball dive off the diving board, you suddenly find yourself bobbing up and down. You can make something move by giving it a push or pull, but the person jumping didn't touch your air mattress. How did the energy from the dive travel through the water and move your air mattress? The up-and-down motion was caused by the peaks and valleys of the ripples that moved from where the splash occurred. These peaks and valleys make up water waves.

Waves Carry Energy Rhythmic disturbances that carry energy without carrying matter are called **waves.** Water waves are shown in **Figure 1.** You can see the energy of the wave from a speedboat traveling outward, but the water only moves up and down. If you've ever felt a clap of thunder, you know that sound waves can carry large amounts of energy. You also transfer energy when you throw something to a friend, as in **Figure 1.** However, there is a difference between a moving ball and a wave. A ball is made of matter, and when it is thrown, the matter moves from one place to another. So, unlike the wave, throwing a ball involves the transport of matter as well as energy.

Figure 1 Water waves and a moving ball both transfer energy from one place to another.

Water waves produced by a boat travel outward, but matter does not move along with the waves.

When the ball is thrown, energy is transferred as matter moves from place to place.

As the students pass the ball, the students' positions do not change—only the position of the ball changes.

In a water wave, water molecules bump each other and pass energy from molecule to molecule.

Figure 2 A wave transports energy without transporting matter from place to place.
Describe *other models that could be used to represent a mechanical wave.*

A Model for Waves

How does a wave carry energy without transporting matter? Imagine a line of people, as shown in **Figure 2.** The first person in line passes a ball to the second person, who passes the ball to the next person, and so on. Passing a ball down a line of people is a model for how waves can transport energy without transporting matter. Even though the ball has traveled, the people in line have not moved. In this model, you can think of the ball as representing energy. What do the people in line represent?

Think about the ripples on the surface of a pond. The energy carried by the ripples travels through the water. The water is made up of water molecules. It is the individual molecules of water that pass the wave energy, just as the people. The water molecules transport the energy in a water wave by colliding with the molecules around them, as shown in **Figure 2.**

Reading Check *What is carried by waves?*

Mechanical Waves

In the wave model, the ball could not be transferred if the line of people didn't exist. The energy of a water wave could not be transferred if no water molecules existed. These types of waves, which use matter to transfer energy, are called **mechanical waves.** The matter through which a mechanical wave travels is called a medium. For ripples on a pond, the medium is the water.

A mechanical wave travels as energy is transferred from particle to particle in the medium. For example, a sound wave is a mechanical wave that can travel through air, as well as solids, liquids, and other gases. Without a medium such as air, there would be no sound waves. In outer space sound waves can't travel because there is no air.

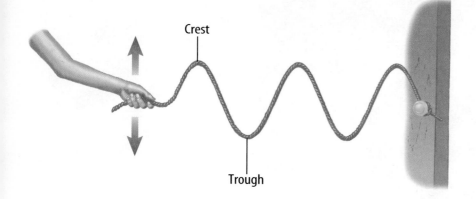

Crest

Trough

Figure 3 The high points on the wave are called crests and the low points are called troughs.

Transverse Waves In a mechanical **transverse wave,** the wave energy causes the matter in the medium to move up and down or back and forth at right angles to the direction the wave travels. You can make a model of a transverse wave. Stretch a long rope out on the ground. Hold one end in your hand. Now shake the end in your hand back and forth. As you shake the rope, you create a wave that seems to slide along the rope.

When you first started shaking the rope, it might have appeared that the rope itself was moving away from you. But it was only the wave that was moving away from your hand. The wave energy moves through the rope, but the matter in the rope doesn't travel. You can see that the wave has peaks and valleys at regular intervals. As shown in **Figure 3,** the high points of transverse waves are called crests. The low points are called troughs.

Reading Check *What are the highest points of transverse waves called?*

Figure 4 A compressional wave can travel through a coiled spring toy.

As the wave motion begins, the coils on the left are close together and the other coils are far apart.

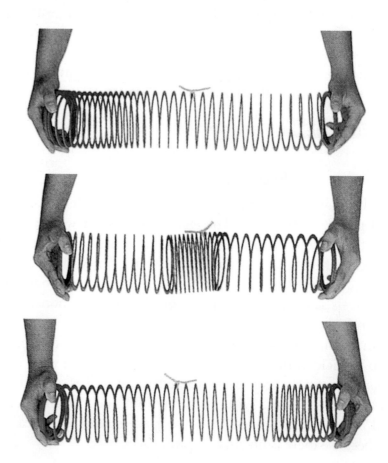

The wave, seen in the squeezed and stretched coils, travels along the spring.

The string and coils did not travel with the wave. Each coil moved forward and then back to its original position.

Compressional Waves Mechanical waves can be either transverse or compressional. In a **compressional wave,** matter in the medium moves forward and backward along the same direction that the wave travels. You can make a compressional wave by squeezing together and releasing several coils of a coiled spring toy, as shown in **Figure 4.**

The coils move only as the wave passes and then return to their original positions. So, like transverse waves, compressional waves carry only energy forward along the spring. In this example, the spring is the medium the wave moves through, but the spring does not move along with the wave.

Sound Waves Sound waves are compressional waves. How do you make sound waves when you talk or sing? If you hold your fingers against your throat while you hum, you can feel vibrations. These vibrations are the movements of your vocal cords. If you touch a stereo speaker while it's playing, you can feel it vibrating, too. All waves are produced by something that is vibrating.

Making Sound Waves

How do vibrating objects make sound waves? Look at the drum shown in **Figure 5.** When you hit the drumhead it starts vibrating up and down. As the drumhead moves upward, the molecules next to it are pushed closer together. This group of molecules that are closer together is a compression. As the compression is formed, it moves away from the drumhead, just as the squeezed coils move along the coiled spring toy in **Figure 4.**

When the drumhead moves downward, the molecules near it have more room and can spread farther apart. This group of molecules that are farther apart is a rarefaction. The rarefaction also moves away from the drumhead. As the drumhead vibrates up and down, it forms a series of compressions and rarefactions that move away and spread out in all directions. This series of compressions and rarefactions is a sound wave.

Comparing Sounds

Procedure
1. Hold a **wooden ruler** firmly on the edge of your **desk** so that most of it extends off the edge of the desk.
2. Pluck the free end of the ruler so that it vibrates up and down. Use gentle motion at first, then pluck with more energy.
3. Repeat step 2, moving the ruler about 1 cm further onto the desk each time until only about 5 cm extend off the edge.

Analysis
1. Compare the loudness of the sounds that are made by plucking the ruler in different ways.
2. Describe the differences in the sound as the end of the ruler extended farther from the desk.

Figure 5 A vibrating drumhead makes compressions and rarefactions in the air.
Describe *how compressions and rarefactions are different.*

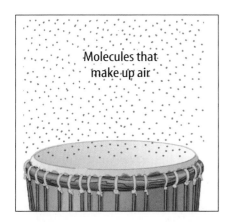

Molecules that make up air

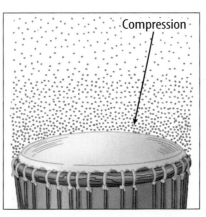

Compression

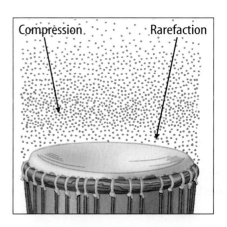

Compression   Rarefaction

Global Positioning Systems
Maybe you've used a global positioning system (GPS) receiver to determine your location while driving, boating, or hiking. Earth-orbiting satellites send electromagnetic radio waves that transmit their exact locations and times of transmission. The GPS receiver uses information from four of these satellites to determine your location to within about 16 m.

Electromagnetic Waves

Waves that can travel through space where there is no matter are **electromagnetic waves.** There are different types of electromagnetic waves, including radio waves, infrared waves, visible light waves, ultraviolet waves, X rays, and gamma rays. These waves can travel in matter or in space. Radio waves from TV and radio stations travel through air, and may be reflected from a satellite in space. They then travel through air, through the walls of your house, and to your TV or radio.

Radiant Energy from the Sun The Sun emits electromagnetic waves that travel through space and reach Earth. The energy carried by electromagnetic waves is called radiant energy. Almost 92 percent of the radiant energy that reaches Earth from the Sun is carried by infrared and visible light waves. Infrared waves make you feel warm when you sit in sunlight, and visible light waves enable you to see. A small amount of the radiant energy that reaches Earth is carried by ultraviolet waves. These are the waves that can cause sunburn if you are exposed to sunlight for too long.

section 1 review

Summary

What is a wave?

- Waves transfer energy, but do not transfer matter.

Mechanical Waves

- Mechanical waves require a medium in which to travel.
- When a transverse wave travels, particles of the medium move at right angles to the direction the wave is traveling.
- When a compressional wave travels, particles of the medium move back and forth along the same direction the wave is traveling.
- Sound is a compressional wave.

Electromagnetic Waves

- Electromagnetic waves can travel through empty space.
- The Sun emits different types of electromagnetic waves, including infrared, visible light, and ultraviolet waves.

Self Check

1. **Describe** the movement of a floating object on a pond when struck by a wave.
2. **Explain** why a sound wave can't travel from a satellite to Earth.
3. **Compare and contrast** a transverse wave and a compressional wave. How are they similar and different?
4. **Compare and contrast** a mechanical wave and an electromagnetic wave.
5. **Think Critically** How is it possible for a sound wave to transmit energy but not matter?

Applying Skills

6. **Concept Map** Create a concept map that shows the relationships among the following: *waves, mechanical waves, electromagnetic waves, compressional waves,* and *transverse waves.*
7. **Use a Word Processor** Use word-processing software to write short descriptions of the waves you encounter during a typical day.

Science online   booko.msscience.com/self_check_quiz

Wave Properties

Amplitude

Can you describe a wave? For a water wave, one way might be to tell how high the wave rises above, or falls below, the normal level. This distance is called the wave's amplitude. The **amplitude** of a transverse wave is one-half the distance between a crest and a trough, as shown in **Figure 6.** In a compressional wave, the amplitude is greater when the particles of the medium are squeezed closer together in each compression and spread farther apart in each rarefaction.

Amplitude and Energy A wave's amplitude is related to the energy that the wave carries. For example, the electromagnetic waves that make up bright light have greater amplitudes than the waves that make up dim light. Waves of bright light carry more energy than the waves that make up dim light. In a similar way, loud sound waves have greater amplitudes than soft sound waves. Loud sounds carry more energy than soft sounds. If a sound is loud enough, it can carry enough energy to damage your hearing.

When a hurricane strikes a coastal area, the resulting water waves carry enough energy to damage almost anything that stands in their path. The large waves caused by a hurricane carry more energy than the small waves or ripples on a pond.

as you read

What You'll Learn

- **Describe** the relationship between the frequency and wavelength of a wave.
- **Explain** why waves travel at different speeds.

Why It's Important

The properties of a wave determine whether the wave is useful or dangerous.

Review Vocabulary

speed: the distance traveled divided by the time needed to travel the distance

New Vocabulary

- amplitude
- wavelength
- frequency

Figure 6 The energy carried by a wave increases as its amplitude increases.

A water wave of large amplitude carried the energy that caused this damage.

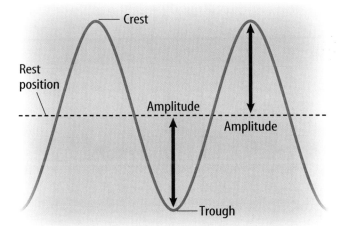

The amplitude of a transverse wave is a measure of how high the crests are or how deep the troughs are.

For transverse waves, wavelength is the distance from crest to crest or trough to trough.

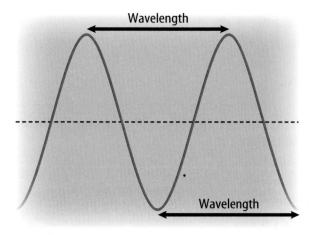
Wavelength

Wavelength

For compressional waves, wavelength is the distance from compression to compression or rarefaction to rarefaction.

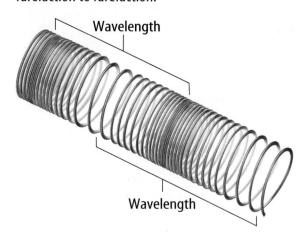
Wavelength

Wavelength

Figure 7 A transverse or a compressional wave has a wavelength.

Figure 8 The wavelengths and frequencies of electromagnetic waves vary.

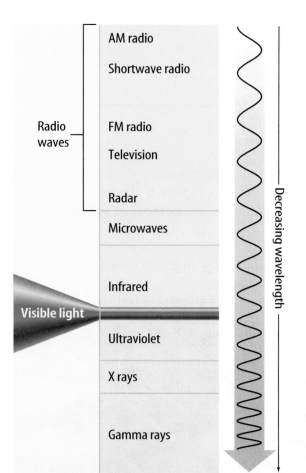

AM radio

Shortwave radio

Radio waves

FM radio

Television

Radar

Microwaves

Infrared

Visible light

Ultraviolet

X rays

Gamma rays

Decreasing wavelength

INTEGRATE
Earth Science

The devastating effect that a wave with large amplitude can have is seen in the aftermath of tsunamis. Tsunamis are huge sea waves that are caused by underwater earthquakes along faults on the seafloor. The movement of the seafloor along a fault produces the wave. As the wave moves toward shallow water and slows down, the amplitude of the wave grows. The tremendous amounts of energy tsunamis carry cause great damage when they move ashore.

Wavelength

Another way to describe a wave is by its wavelength. **Figure 7** shows the wavelength of a transverse wave and a compressional wave. For a transverse wave, **wavelength** is the distance from the top of one crest to the top of the next crest, or from the bottom of one trough to the bottom of the next trough. For a compressional wave, the wavelength is the distance between the center of one compression and the center of the next compression, or from the center of one rarefaction to the center of the next rarefaction.

Electromagnetic waves have wavelengths that range from kilometers, for radio waves, to less than the diameter of an atom, for X rays and gamma rays. This range is called the electromagnetic spectrum. **Figure 8** shows the names given to different parts of the electromagnetic spectrum. Visible light is only a small part of the electromagnetic spectrum. It is the wavelength of visible light waves that determines their color. For example, the wavelength of red light waves is longer than the wavelength of green light waves.

Frequency

The **frequency** of a wave is the number of wavelengths that pass a given point in 1 s. The unit of frequency is the number of wavelengths per second, or hertz (Hz). Recall that waves are produced by something that vibrates. The faster the vibration is, the higher the frequency is of the wave that is produced.

Reading Check *How is the frequency of a wave measured?*

A Sidewalk Model For waves that travel with the same speed, frequency and wavelength are related. To model this relationship, imagine people on two parallel moving sidewalks in an airport, as shown in **Figure 9.** One sidewalk has four travelers spaced 4 m apart. The other sidewalk has 16 travelers spaced 1 m apart.

Now imagine that both sidewalks are moving at the same speed and approaching a pillar between them. On which sidewalk will more people go past the pillar? On the sidewalk with the shorter distance between people, four people will pass the pillar for each one person on the other sidewalk. When four people pass the pillar on the first sidewalk, 16 people pass the pillar on the second sidewalk.

Figure 9 When people are farther apart on a moving sidewalk, fewer people pass the pillar every minute.
Infer *how the number of people passing the pillar each minute would change if the sidewalk moved slower.*

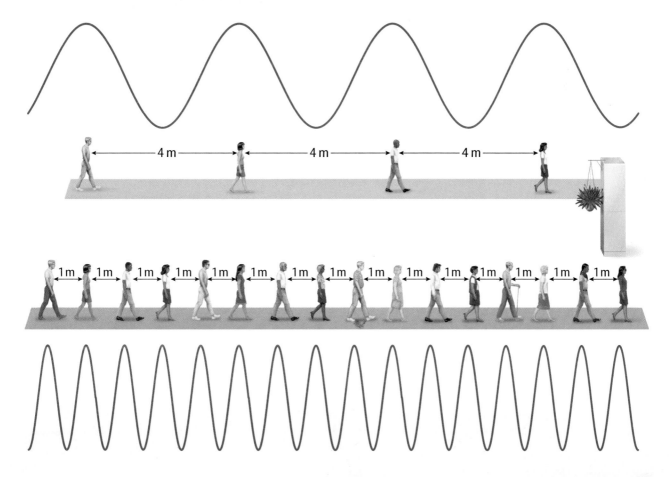

Ultrasonic Waves Sound waves with ultra-high frequencies cannot be heard by the human ear, but they are used by medical professionals in several ways. They are used to perform echocardiograms of the heart, produce ultrasound images of internal organs, break up blockages in arteries, and sterilize surgical instruments. Describe how the wavelengths of these sound waves compare to sound waves you can hear.

Figure 10 The frequency of the notes on a musical scale increases as the notes get higher in pitch, but the wavelength of the notes decreases.

Frequency and Wavelength Suppose that each person in **Figure 9** represents the crest of a wave. Then the movement of people on the first sidewalk is like a wave with a wavelength of 4 m. For the second sidewalk, the wavelength would be 1 m. On the first sidewalk, where the wavelength is longer, the people pass the pillar *less* frequently. Smaller frequencies result in longer wavelengths. On the second sidewalk, where the wavelength is shorter, the people pass the pillar *more* frequently. Higher frequencies result in shorter wavelengths. This is true for all waves that travel at the same speed. As the frequency of a wave increases, its wavelength decreases.

Reading Check *How are frequency and wavelength related?*

Color and Pitch Because frequency and wavelength are related, either the wavelength or frequency of a light wave determines the color of the light. For example, blue light has a larger frequency and shorter wavelength than red light.

Either the wavelength or frequency determines the pitch of a sound wave. Pitch is how high or low a sound seems to be. When you sing a musical scale, the pitch and frequency increase from note to note. Wavelength and frequency are also related for sound waves traveling in air. As the frequency of sound waves increases, their wavelength decreases. **Figure 10** shows how the frequency and wavelength change for notes on a musical scale.

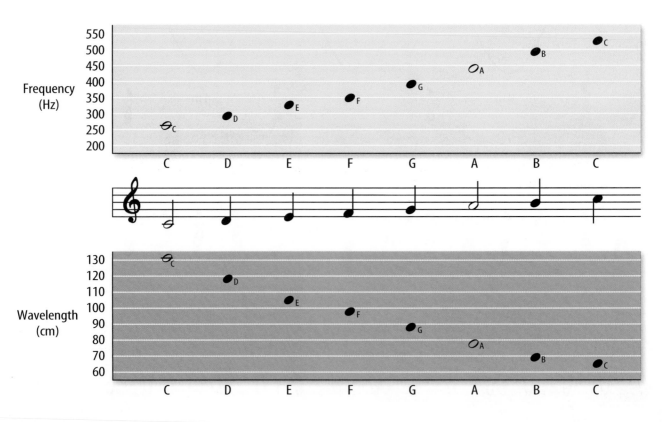

Wave Speed

You've probably watched a distant thunderstorm approach on a hot summer day. You see a bolt of lightning flash between a dark cloud and the ground. If the thunderstorm is many kilometers away, several seconds will pass between when you see the lightning and when you hear the thunder. This happens because light travels much faster in air than sound does. Light travels through air at about 300 million m/s. Sound travels through air at about 340 m/s. The speed of any wave can be calculated from this equation:

Wave Speed Equation

wave speed (in m/s) = **frequency** (in Hz) × **wavelength** (m)

$$v = f\lambda$$

In this equation, the wavelength is represented by the symbol λ, which is the Greek letter lambda.

When mechanical waves, such as sound, and electromagnetic waves, such as light, travel in different materials, they change speed. Mechanical waves usually travel fastest in solids, and slowest in gases. Electromagnetic waves travel fastest in gases and slowest in solids. For example, the speed of light is about 30 percent faster in air than in water.

Science online

Topic: Wave Speed
Visit booko.msscience.com for Web links to information about wave speed in different materials.

Activity Make a chart showing the speed of light in different materials.

section 2 review

Summary

Amplitude

- In a transverse wave, the amplitude is one-half the distance between a crest and a trough.
- The larger the amplitude, the greater the energy carried by the wave.

Wavelength

- For a transverse wave, wavelength is the distance from crest to crest, or from trough to trough.
- For a compressional wave, wavelength is the distance from compression to compression, or from rarefaction to rarefaction.

Frequency

- The frequency of a wave is the number of wavelengths that pass a given point in 1 s.
- For waves that travel at the same speed, as the frequency of the wave increases, its wavelength decreases.

Self Check

1. **Describe** how the frequency of a wave changes as its wavelength changes.
2. **Explain** why a sound wave with a large amplitude is more likely to damage your hearing than one with a small amplitude.
3. **State** what accounts for the time difference between seeing and hearing a fireworks display.
4. **Explain** why the statement "The speed of light is 300 million m/s" is not always correct.
5. **Think Critically** Explain the differences between the waves that make up bright, green light and dim, red light.

Applying Math

6. **Calculate Wave Speed** Find the speed of a wave with a wavelength of 5 m and a frequency of 68 Hz.
7. **Calculate Wavelength** Find the wavelength of a sound wave traveling in water with a speed of 1,470 m/s, and a frequency of 2,340 Hz.

Waves on a Spring

Waves are rhythmic disturbances that carry energy through matter or space. Studying waves can help you understand how the Sun's energy reaches Earth and sounds travel through the air.

◉ Real-World Question

What are some of the properties of transverse and compressional waves on a coiled spring?

Goals

- **Create** transverse and compressional waves on a coiled spring toy.
- **Investigate** wave properties such as speed and amplitude.

Materials

long, coiled spring toy
colored yarn (5 cm)
meterstick
stopwatch

Safety Precautions 🥽 🔥

WARNING: *Avoid overstretching or tangling the spring to prevent injury or damage.*

◉ Procedure

1. Prepare a data table such as the one shown.

Wave Data	
Length of stretched spring toy	Do not write in this book.
Average time for a wave to travel from end to end—step 4	
Average time for a wave to travel from end to end—step 5	

2. Work in pairs or groups and clear a place on an uncarpeted floor about 6 m × 2 m.

3. Stretch the springs between two people to the length suggested by your teacher. Measure the length.

4. Create a wave with a quick, sideways snap of the wrist. Time several waves as they travel the length of the spring. Record the average time in your data table.

5. Repeat step 4 using waves that have slightly larger amplitudes.

6. Squeeze together about 20 of the coils. Observe what happens to the unsqueezed coils. Release the coils and observe.

7. Quickly push the spring toward your partner, then pull it back.

8. Tie the yarn to a coil near the middle of the spring. Repeat step 7, observing the string.

9. **Calculate** and compare the speeds of the waves in steps 4 and 5.

◉ Conclude and Apply

1. **Classify** the wave pulses you created in each step as compressional or transverse.

2. **Classify** the unsqueezed coils in step 6 as a compression or a rarefaction.

3. **Compare and contrast** the motion of the yarn in step 8 with the motion of the wave.

𝒞ommunicating Your Data

Write a summary paragraph of how this lab demonstrated any of the vocabulary words from the first two sections of the chapter. **For more help, refer to the** Science Skill Handbook.

(3)

Wave Behavior

Reflection

What causes the echo when you yell across an empty gymnasium or down a long, empty hallway? Why can you see your face when you look in a mirror? The echo of your voice and the face you see in the mirror are caused by wave reflection.

Reflection occurs when a wave strikes an object or surface and bounces off. An echo is reflected sound. Sound reflects from all surfaces. Your echo bounces off the walls, floor, ceiling, furniture, and people. You see your face in a mirror or a still pond, as shown in **Figure 11,** because of reflection. Light waves produced by a source of light such as the Sun or a lightbulb bounce off your face, strike the mirror, and reflect back to your eyes.

When a surface is smooth and even the reflected image is clear and sharp. However, **Figure 11** shows that when light reflects from an uneven or rough surface, you can't see a sharp image because the reflected light scatters in many different directions.

Reading Check *What causes reflection?*

What **You'll Learn**

- **Explain** how waves can reflect from some surfaces.
- **Explain** how waves change direction when they move from one material into another.
- **Describe** how waves are able to bend around barriers.

Why **It's Important**

The reflection of waves enables you to see objects around you.

Review Vocabulary
echo: the repetition of a sound caused by the reflection of sound waves

New Vocabulary
- reflection
- refraction
- diffraction
- interference

Figure 11 The image formed by reflection depends on the smoothness of the surface.

The smooth surface of a still pond enables you to see a sharp, clear image of yourself.

If the surface of the pond is rough and uneven, your reflected image is no longer clear and sharp.

SECTION 3 Wave Behavior **0** ◆ **19**

Mini LAB

Observing How Light Refracts

Procedure 🥽

1. Fill a **large, opaque drinking glass or cup** with **water.**
2. Place a **white soda straw** in the water at an angle.
3. Looking directly down into the cup from above, observe the straw where it meets the water.
4. Placing yourself so that the straw angles to your left or right, slowly back away about 1 m. Observe the straw as it appears above, at, and below the surface of the water.

Analysis

1. Describe the straw's appearance from above.
2. Compare the straw's appearance above and below the water's surface in step 4.

Try at Home

Refraction

A wave changes direction when it reflects from a surface. Waves also can change direction in another way. Perhaps you have tried to grab a sinking object when you are in a swimming pool, only to come up empty-handed. Yet you were sure you grabbed right where you saw the object. You missed grabbing the object because the light rays from the object changed direction as they passed from the water into the air. The bending of a wave as it moves from one medium into another is called **refraction.**

Refraction and Wave Speed Remember that the speed of a wave can be different in different materials. For example, light waves travel faster in air than in water. Refraction occurs when the speed of a wave changes as it passes from one substance to another, as shown in **Figure 12.** A line that is perpendicular to the water's surface is called the normal. When a light ray passes from air into water, it slows down and bends toward the normal. When the ray passes from water into air, it speeds up and bends away from the normal. The larger the change in speed of the light wave is, the larger the change in direction is.

You notice refraction when you look down into a fishbowl. Refraction makes the fish appear to be closer to the surface and farther away from you than it really is, as shown in **Figure 13.** Light rays reflected from the fish are bent away from the normal as they pass from water to air. Your brain interprets the light that enters your eyes by assuming that light rays always travel in straight lines. As a result, the light rays seem to be coming from a fish that is closer to the surface.

Figure 12 A wave is refracted when it changes speed.

Explain *how the direction of the light ray changes if it doesn't change speed.*

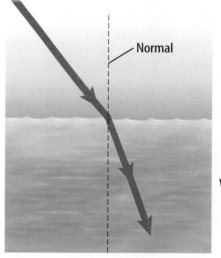

As the light ray passes from air to water, it bends toward the normal.

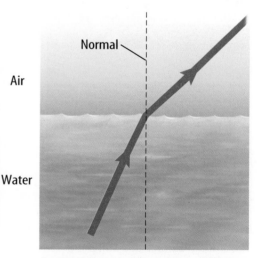

As the light ray passes from water to air, it bends away from the normal.

Within the Figure 12 diagram: Normal, Air, Water

Color from Refraction Sunlight contains light of various wavelengths. When sunlight passes through a prism, refraction occurs twice: once when sunlight enters the prism and again when it leaves the prism and returns to the air. Violet light has the shortest wavelength and is bent the most. Red light has the longest wavelength and is bent the least. Each color has a different wavelength and is refracted a different amount. As a result, the colors of sunlight are separated when they emerge from the prism.

Figure 14 shows how refraction produces a rainbow when light waves from the Sun pass into and out of water droplets. The colors you see in a rainbow are in order of decreasing wavelength: red, orange, yellow, green, blue, indigo, and violet.

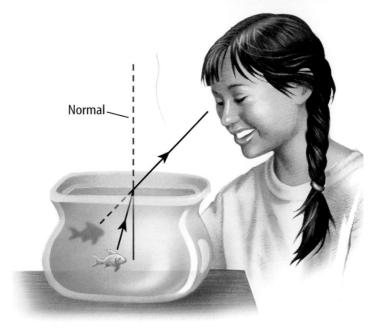

Figure 13 When you look at the goldfish in the water, the fish is in a different position than it appears.
Infer *how the location of the fish would change if light traveled faster in water than in air.*

Diffraction

Why can you hear music from the band room when you are down the hall? You can hear the music because the sound waves bend as they pass through an open doorway. This bending isn't caused by refraction. Instead, the bending is caused by diffraction. **Diffraction** is the bending of waves around a barrier.

Light waves can diffract, too. You can hear your friends in the band room but you can't see them until you reach the open door. Therefore, you know that light waves do not diffract as much as sound waves do. Light waves do bend around the edges of an open door. However, for an opening as wide as a door, the amount the light bends is extremely small. As a result, the diffraction of light is far too small to allow you to see around a corner.

Figure 14 Light rays refract as they enter and leave each water drop. Each color refracts at different angles because of their different wavelengths, so they separate into the colors of the visible spectrum.

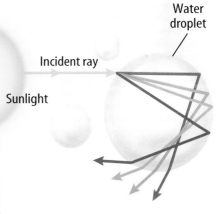

Water droplet

Incident ray

Sunlight

Diffraction and Wavelength The reason that light waves don't diffract much when they pass through an open door is that the wavelengths of visible light are much smaller than the width of the door. Light waves have wavelengths between about 400 and 700 billionths of a meter, while the width of doorway is about one meter. Sound waves that you can hear have wavelengths between a few millimeters and about 10 m. They bend more easily around the corners of an open door. A wave is diffracted more when its wavelength is similar in size to the barrier or opening.

✓ Reading Check *Under what conditions would more diffraction of a wave occur?*

Diffraction of Water Waves Perhaps you have noticed water waves bending around barriers. For example, when water waves strike obstacles such as the islands shown in **Figure 15,** they don't stop moving. Here the size and spacing of the islands is not too different from the wavelength of the water waves. So the water waves bend around the islands, and keep on moving. They also spread out after they pass through openings between the islands. If the islands were much larger than the water wavelength, less diffraction would occur.

What happens when waves meet?

Suppose you throw two pebbles into a still pond. Ripples spread from the impact of each pebble and travel toward each other. What happens when two of these ripples meet? Do they collide like billiard balls and change direction? Waves behave differently from billiard balls when they meet. Waves pass right through each other and continue moving.

Figure 15 Water waves bend or diffract around these islands. More diffraction occurs when the object is closer in size to the wavelength.

Wave Interference While two waves overlap a new wave is formed by adding the two waves together. The ability of two waves to combine and form a new wave when they overlap is called **interference.** After they overlap, the individual waves continue to travel on in their original form.

The different ways waves can interfere are shown in **Figure 16** on the next page. Sometimes when the waves meet, the crest of one wave overlaps the crest of another wave. This is called constructive interference. The amplitudes of these combining waves add together to make a larger wave while they overlap. Destructive interference occurs when the crest of one wave overlaps the trough of another wave. Then, the amplitudes of the two waves combine to make a wave with a smaller amplitude. If the two waves have equal amplitudes and meet crest to trough, they cancel each other while the waves overlap.

Waves and Particles Like waves of water, when light travels through a small opening, such as a narrow slit, the light spreads out in all directions on the other side of the slit. If small particles, instead of waves, were sent through the slit, they would continue in a straight line without spreading. The spreading, or diffraction, is only a property of waves. Interference also doesn't occur with particles. If waves meet, they reinforce or cancel each other, then travel on. If particles approach each other, they either collide and scatter or miss each other completely. Interference, like diffraction, is a property of waves.

Science nline

Topic: Interference
Visit booko.msscience.com for Web links to information about wave interference.

Activity Write a paragraph about three kinds of interference you found in your research.

Applying Science

Can you create destructive interference?

Your brother is vacuuming and you can't hear the television. Is it possible to diminish the sound of the vacuum so you can hear the TV? Can you eliminate some sound waves and keep the sounds you do want to hear?

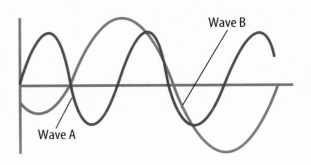

Identifying the Problem

It is possible to create a wave that will destructively interfere with one wave, but will not destructively interfere with another wave. The graph shows two waves with different wavelengths.

Solving the Problem

1. Create the graph of a wave that will eliminate wave **A** but not wave **B**.
2. Create the graph of a wave that would amplify wave **A**.

Figure 16

Whether they are ripples on a pond or huge ocean swells, when water waves meet they can combine to form new waves in a process called interference. As shown below, wave interference can be constructive or destructive.

Constructive Interference

In constructive interference, a wave with greater amplitude is formed.

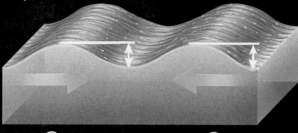

The crests of two waves—A and B—approach each other.

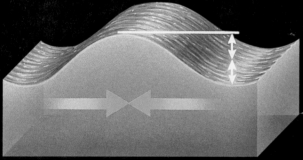

The two waves form a wave with a greater amplitude while the crests of both waves overlap.

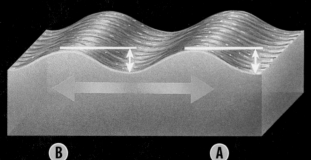

The original waves pass through each other and go on as they started.

Destructive Interference

In destructive interference, a wave with a smaller amplitude is formed.

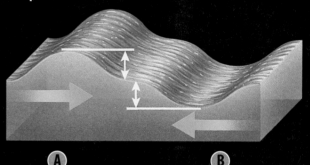

The crest of one wave approaches the trough of another.

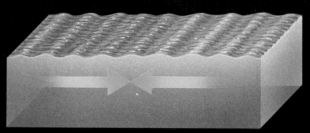

If the two waves have equal amplitude, they momentarily cancel when they meet.

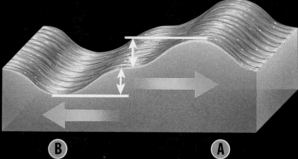

The original waves pass through each other and go on as they started.

Reducing Noise You might have seen someone use a power lawn mower or a chain saw. In the past, many people who performed these tasks damaged their hearing because of the loud noises produced by these machines.

Loud sounds have waves with larger amplitudes and carry more energy than softer sounds. The energy carried by loud sounds can damage cells in the ear that vibrate and transmit signals to the brain. Damage to the ear from loud sounds can be prevented by reducing the energy that reaches the ear. Ear protectors contain materials that absorb some of the energy carried by sound waves, so that less sound energy reaches the ear.

Pilots of small planes have a more complicated problem. If they shut out all the noise of the plane's motor, the pilots wouldn't be able to hear instructions from air-traffic controllers. To solve this problem, ear protectors have been developed, as shown in **Figure 17,** that have electronic circuits. These circuits detect noise from the aircraft and produce sound frequencies that destructively interfere with the noise. They do not interfere with human voices, so people can hear normal conversation. Destructive interference can be a benefit.

Figure 17 Some airplane pilots use special ear protectors that cancel out engine noise but don't block human voices.

section 3 review

Summary

Reflection
- Reflected sound waves can produce echoes.
- Reflected light rays produce images in a mirror.

Refraction
- The bending of waves as they pass from one medium to another is refraction.
- Refraction occurs when the wave's speed changes.
- A prism separates sunlight into the colors of the visible spectrum.

Diffraction and Interference
- The bending of waves around barriers is diffraction.
- Interference occurs when waves combine to form a new wave while they overlap.
- Destructive interference can reduce noise.

Self Check

1. **Explain** why you don't see your reflection in a building made of rough, white stone.
2. **Explain** how you are able to hear the siren of an ambulance on the other side of a building.
3. **Describe** the behavior of light that enables magnifying lenses and contact lenses to bend light rays.
4. **Define** the term *diffraction*. How does the amount of diffraction depend on wavelength?
5. **Think Critically** Why don't light rays that stream through an open window into a darkened room spread evenly through the entire room?

Applying Skills

6. **Compare and Contrast** When light rays pass from water into a certain type of glass, the rays refract toward the normal. Compare and contrast the speed of light in water and in the glass.

Design Your Own

WAVE ≈ SPEED

Goals
- **Measure** the speed of a wave within a coiled spring toy.
- **Predict** whether the speed you measured will be different in other types of coiled spring toys.

Possible Materials
long, coiled spring toy
meterstick
stopwatch
tape
*clock with a second hand
*Alternate materials

Safety Precautions

⊙ Real-World Question

When an earthquake occurs, it produces waves that are recorded at points all over the world by instruments called seismographs. By comparing the data that they collected from these seismographs, scientists discovered that the interior of Earth must be made of layers of different materials. These data showed that the waves traveled at different speeds as they passed through different parts of Earth's interior. How can the speed of a wave be measured?

⊙ Form a Hypothesis

In some materials, waves travel too fast for their speeds to be measured directly. Think about what you know about the relationships among the frequency, wavelength, and speed of a wave in a medium. Make a hypothesis about how you can use this relationship to measure the speed of a wave within a medium. Explain why you think the experiment will support your hypothesis.

⊙ Test Your Hypothesis

Make a Plan

1. Make a data table in your Science Journal like the one shown.

2. In your Science Journal, write a detailed description of the coiled spring toy you are going to use. Be sure to include its mass and diameter, the width of a coil, and what it is made of.

3. **Decide** as a group how you will measure the frequency and length of waves in the spring toy. What are your variables? Which variables must be controlled? What variable do you want to measure?

4. Repeat your experiment three times.

Follow Your Plan

1. Make sure your teacher approves your plan before you start.
2. Carry out the experiment.
3. While you are doing the experiment, record your observations and measurements in your data table.

Wave Data			
	Trial 1	Trial 2	Trial 3
Length spring was stretched (m)			
Number of crests			
Wavelength (m)			
# of vibrations timed	Do not write in this book.		
# of seconds vibrations were timed			
Wave speed (m/s)			

▶ Analyze Your Data

1. **Calculate** the frequency of the waves by dividing the number of vibrations you timed by the number of seconds you timed them. Record your results in your data table.

2. Use the following formula to calculate the speed of a wave in each trial.

   wavelength $\times$ frequency $=$ wave speed

3. Average the wave speeds from your trials to determine the average speed of a wave in your coiled spring toy.

▶ Conclude and Apply

1. **Infer** which variables affected the wave speed in spring toys the most. Which variables affected the speed the least? Was your hypothesis supported?

2. **Analyze** what factors caused the wave speed measured in each trial to be different.

Communicating Your Data

Post a description of your coiled spring toy and the results of your experiment on a bulletin board in your classroom. **Compare and contrast** your results with other students in your class.

SCIENCE Stats

Waves, Waves, and More Waves

Did you know...

. . . Radio waves from space were discovered in 1932 by Karl G. Jansky, an American engineer. His discovery led to the creation of radio astronomy, a field that explores parts of the universe that can't be seen with telescopes.

. . . The highest recorded ocean wave was 34 meters high, which is comparable to the height of a ten-story building. This super wave was seen in the North Pacific Ocean and recorded by the crew of the naval ship *USS Ramapo* in 1933.

Applying Math   A tsunami formed by an earthquake on the ocean floor travels at 900 km/h. How long will it take the tsunami to travel 4,500 km?

. . . Waves let dolphins see with their ears! A dolphin sends out ultrasonic pulses, or clicks, at rates of 800 pulses per second. These sound waves are reflected back to the dolphin after they hit an obstacle or a meal. This process is called echolocation.

Graph It

Go to booko.msscience.com/science_stats to learn about discoveries by radio astronomers. Make a time line showing some of these discoveries.

Reviewing Main Ideas

Section 1 What are waves?

1. Waves are rhythmic disturbances that carry energy but not matter.

2. Mechanical waves can travel only through matter. Electromagnetic waves can travel through matter and space.

3. In a mechanical transverse wave, matter in the medium moves back and forth at right angles to the direction the wave travels.

4. In a compressional wave, matter in the medium moves forward and backward in the same direction as the wave.

Section 2 Wave Properties

1. The amplitude of a transverse wave is the distance between the rest position and a crest or a trough.

2. The energy carried by a wave increases as the amplitude increases.

3. Wavelength is the distance between neighboring crests or neighboring troughs.

4. The frequency of a wave is the number of wavelengths that pass a given point in 1 s.

5. Waves travel through different materials at different speeds.

Section 3 Wave Behavior

1. Reflection occurs when a wave strikes an object or surface and bounces off.

2. The bending of a wave as it moves from one medium into another is called refraction. A wave changes direction, or refracts, when the speed of the wave changes.

3. The bending of waves around a barrier is called diffraction.

4. Interference occurs when two or more waves combine and form a new wave while they overlap.

Visualizing Main Ideas

Copy and complete the following spider map about waves.

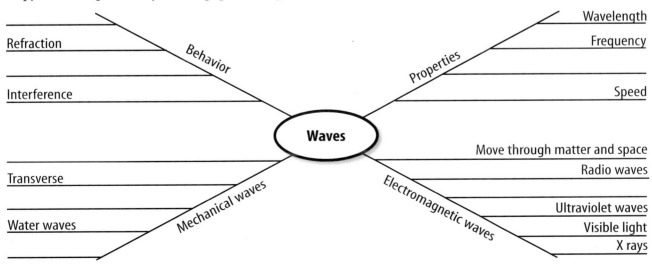

Using Vocabulary

amplitude p. 13	interference p. 23
compressional	mechanical wave p. 9
wave p. 11	reflection p. 19
diffraction p. 21	refraction p. 20
electromagnetic	transverse wave p. 10
wave p. 12	wave p. 8
frequency p. 15	wavelength p. 14

Fill in the blanks with the correct word or words.

1. _____ is the change in direction of a wave going from one medium to another.

2. The type of wave that has rarefactions is a _____.

3. The distance between two adjacent crests of a transverse wave is the _____.

4. The more energy a wave carries, the greater its _____ is.

5. A(n) _____ can travel through space without a medium.

Checking Concepts

Choose the word or phrase that best answers the question.

6. What is the material through which mechanical waves travel?
   A) charged particles
   B) space
   C) a vacuum
   D) a medium

7. What is carried from particle to particle in a water wave?
   A) speed   C) energy
   B) amplitude   D) matter

8. What are the lowest points on a transverse wave called?
   A) crests   C) compressions
   B) troughs   D) rarefactions

9. What determines the pitch of a sound wave?
   A) amplitude   C) speed
   B) frequency   D) refraction

10. What is the distance between adjacent wave compressions?
    A) one wavelength
    B) 1 km
    C) 1 m/s
    D) 1 Hz

11. What occurs when a wave strikes an object or surface and bounces off?
    A) diffraction
    B) refraction
    C) a transverse wave
    D) reflection

12. What is the name for a change in the direction of a wave when it passes from one medium into another?
    A) refraction   C) reflection
    B) interference   D) diffraction

Use the figure below to answer question 13.

13. What type of wave is a sound wave?
    A) transverse
    B) electromagnetic
    C) compressional
    D) refracted

14. What color light has the shortest wavelength and the highest frequency?
    A) red   C) Orange
    B) green   D) Blue

Science Online booko.msscience.com/vocabulary_puzzlemaker

Thinking Critically

15. Explain what kind of wave—transverse or compressional—is produced when an engine bumps into a string of coupled railroad cars on a track.

16. Infer Is it possible for an electromagnetic wave to travel through a vacuum? Through matter? Explain your answers.

17. Draw a Conclusion Why does the frequency of a wave decrease as the wavelength increases?

18. Explain why you don't see your reflected image when you look at a white, rough surface?

19. Infer If a cannon fires at a great distance from you, why do you see the flash before you hear the sound?

20. Form a Hypothesis Form a hypothesis that can explain this observation. Waves A and B travel away from Earth through Earth's atmosphere. Wave A continues on into space, but wave B does not.

Use the figure below to answer question 21.

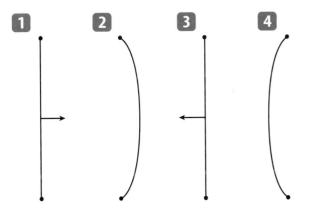

21. Explain how the object shown above causes compressions and rarefactions as it vibrates in air.

22. Explain why you can hear a person talking even if you can't see them.

23. Compare and Contrast AM radio waves have wavelengths between about 200 m and 600 m, and FM radio waves have wavelengths of about 3 m. Why can AM radio signals often be heard behind buildings and mountains but FM radio signals cannot?

24. Infer how the wavelength of a wave would change if the speed of the wave increased, but the frequency remained the same.

25. Explain You are motionless on a rubber raft in the middle of a pool. A friend sitting on the edge of the pool tries to make the float move to the other edge of the pool by slapping the water every second to form a wave. Explain whether the wave produced will cause you to move to the edge of the pool.

Performance Activities

26. Make Flashcards Work with a partner to make flashcards for the bold-faced terms in the chapter. Illustrate each term on the front of the cards. Write the term and its definition on the back of the card. Use the cards to review the terms with another team.

Applying Math

Use the following equation to answer questions 27–29.

wave speed = wavelength × frequency

27. Wave Speed If a wave pool generates waves with a wavelength of 3.2 m and a frequency of 0.60 Hz, how fast are the waves moving?

28. Frequency An earthquake wave travels at 5000 m/s and has a wavelength of 417 m. What is its frequency?

29. Wavelength A wave travels at a velocity of 4 m/s. It has a frequency of 3.5 Hz. What is the wavelength of the wave?

Part 1 | Multiple Choice

Record your answers on the answer sheet provided by your teacher or on a sheet of paper.

1. What do waves carry as they move?
   - **A.** matter
   - **B.** energy
   - **C.** matter and energy
   - **D.** particles and energy

Use the figure below to answer questions 2 and 3.

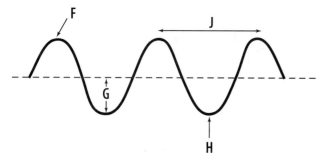

2. What property of the wave is shown at F?
   - **A.** amplitude
   - **B.** wavelength
   - **C.** crest
   - **D.** trough

3. What property of the wave is shown at J?
   - **A.** amplitude
   - **B.** wavelength
   - **C.** crest
   - **D.** trough

4. What kind of wave does NOT need a medium through which to travel?
   - **A.** mechanical
   - **B.** sound
   - **C.** light
   - **D.** refracted

5. What happens as a sound wave's energy decreases?
   - **A.** Wave frequency decreases.
   - **B.** Wavelength decreases.
   - **C.** Amplitude decreases.
   - **D.** Wave speed decreases.

6. What unit is used to measure frequency?
   - **A.** meters
   - **B.** meters/second
   - **C.** decibels
   - **D.** hertz

7. What properties of a light wave determines its color?
   - **A.** wavelength
   - **B.** amplitude
   - **C.** speed
   - **D.** interference

8. When two waves overlap and interfere constructively, what does the resulting wave have?
   - **A.** a greater amplitude
   - **B.** less energy
   - **C.** a change in frequency
   - **D.** a lower amplitude

9. What happens when light travels from air into glass?
   - **A.** It speeds up.
   - **B.** It slows down.
   - **C.** It travels at 300,000 km/s.
   - **D.** It travels at the speed of sound.

Use the figure below to answer questions 10 and 11.

10. What behavior of light waves lets you see a sharp, clear image of yourself?
    - **A.** refraction
    - **B.** diffraction
    - **C.** reflection
    - **D.** interference

11. Why can't you see a clear image of yourself if the water's surface is rough?
    - **A.** The light bounces off the surface in only one direction.
    - **B.** The light scatters in many different directions.
    - **C.** There is no light shining on the water's surface.
    - **D.** The light changes speed when it strikes the water.

Part 2 | Short Response/Grid In

Record your answers on the answer sheet provided by your teacher or on a sheet of paper.

12. An earthquake in the middle of the Indian Ocean produces a tsunami that hits an island. Is the water that hits the island the same water that was above the place where the earthquake occurred? Explain.

Use the figure below to answer questions 13 and 14.

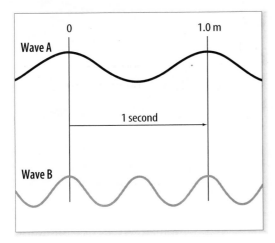

13. Compare the wavelengths and frequencies of the two waves shown.

14. If both waves are traveling through the same medium, how do their speeds compare? Explain.

15. Suppose you make waves in a pond by dipping your hand in the water with a frequency of 1 Hz. How could you make waves of a longer wavelength? How could you increase the amplitude of the waves?

16. How are all electromagnetic waves alike? How do they differ from one another?

Test-Taking Tip

Take Your Time Stay focused during the test and don't rush, even if you notice that other students are finishing the test early.

Part 3 | Open Ended

Record your answers on a sheet of paper.

Use the figure below to answer questions 17 and 18.

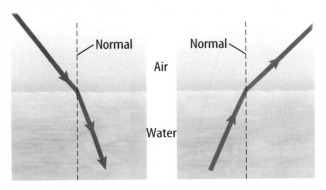

17. Why does the light ray bend toward the normal when is passes from air into water, but bend away from the normal as it passes from water into air?

18. A boy has caught a fish on his fishing line. He reels the fish in near the boat. How could the refraction of light waves affect him as he tries to net the fish while it is still in the water?

19. In a science fiction movie, a spaceship explodes. The people in a nearby spaceship see and hear the explosion. Is this realistic? Explain.

20. The speed of light in warm air is greater than its speed in cold air. The air just above a highway is warmer than the air a little higher. Will the light moving parallel to the highway be bent up or down? Explain.

21. You are standing outside a classroom with an open door. You know your friends are in the room because you can hear them talking. Explain why you can hear them talking but cannot see them.

22. How does the size of an obstacle affect the diffraction of a wave?

Sound

The BIG Idea

Sound waves are compressional waves produced by something that vibrates.

SECTION 1
What is sound?

Main Idea A vibrating object produces compressions and rarefactions that travel away from the object.

SECTION 2
Music

Main Idea The sound of a musical instrument depends on the natural frequencies of the materials it is made from.

An Eerie Silence

You probably have never experienced complete silence unless you've been in a room like this one. The room is lined with special materials that absorb sound waves and eliminate sound reflections.

Science Journal Write a paragraph about the quietest place you've ever been.

Start-Up Activities

Making Human Sounds

When you speak or sing, you push air from your lungs past your vocal cords, which are two flaps of tissue inside your throat. When you tighten your vocal cords, you can make the sound have a higher pitch. Do this lab to explore how you change the shape of your throat to vary the pitch of sound.

1. Hold your fingers against the front of your throat and say *Aaaah*. Notice the vibration against your fingers.
2. Now vary the pitch of this sound from low to high and back again. How do the vibrations in your throat change? Record your observations.
3. Change the sound to an *Ooooh*. What do you notice as you listen? Record your observations.
4. **Think Critically** In your Science Journal, describe how the shape of your throat changed the pitch.

 Preview this chapter's content and activities at booko.msscience.com

Sound Make the following Foldable to help you answer questions about sound.

STEP 1 Fold a vertical sheet of notebook paper from side to side.

STEP 2 Cut along every third line of only the top layer to form tabs.

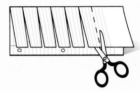

STEP 3 Write a question about sound on each tab.

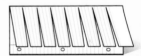

Answer Questions Before you read the chapter, write some questions you have about sound on the front of the tabs. As you read the chapter, write the answer beneath the question. You may add questions as you read.

Get Ready to Read

Monitor

① Learn It! An important strategy to help you improve your reading is monitoring, or finding your reading strengths and weaknesses. As you read, monitor yourself to make sure the text makes sense. Discover different monitoring techniques you can use at different times, depending on the type of test and situation.

② Practice It! The paragraph below appears in Section 1. Read the passage and answer the questions that follow. Discuss your answers with other students to see how they monitor their reading.

> The diffraction of lower frequencies in the human voice allows you to hear someone talking even when the person is around the corner. This is different from an echo. Echoes occur when sound waves bounce off a reflecting surface. Diffraction occurs when a wave spreads out after passing through an opening or when a wave bends around an obstacle.
>
> — *from page 44*

- What questions do you still have after reading?
- Do you understand all of the words in the passage?
- Did you have to stop reading often? Is the reading level appropriate for you?

③ Apply It! Identify one paragraph that is difficult to understand. Discuss it with a partner to improve your understanding.

Reading Tip

Monitor your reading by slowing down or speeding up, depending on your understanding of the text.

Target Your Reading

Use this to focus on the main ideas as you read the chapter.

1 **Before you read** the chapter, respond to the statements below on your worksheet or on a numbered sheet of paper.
- Write an **A** if you **agree** with the statement.
- Write a **D** if you **disagree** with the statement.

2 **After you read** the chapter, look back to this page to see if you've changed your mind about any of the statements.
- If any of your answers changed, explain why.
- Change any false statements into true statements.
- Use your revised statements as a study guide.

Science Online
Print out a worksheet of this page at booko.msscience.com

Before You Read A or D		Statement	After You Read A or D
	1	Sound waves transfer energy only in matter.	
	2	The loudness of a sound wave increases as the frequency of a wave increases.	
	3	Sound travels faster in warm air than in cold air.	
	4	Sound usually travels faster in gases than in solids.	
	5	The pitch of a sound you hear depends on whether the source of the sound is moving relative to you.	
	6	Sound waves do not spread out when they pass through an opening.	
	7	A vibrating string whose length is fixed can produce sound waves of more than one frequency.	
	8	The body of a guitar helps make the sound of the vibrating strings louder.	
	9	Changing the length of a vibrating air column changes the pitch of the sound produced.	

What is sound?

What **You'll Learn**

- **Identify** the characteristics of sound waves.
- **Explain** how sound travels.
- **Describe** the Doppler effect.

Why **It's Important**

Sound gives important information about the world around you.

⊙ **Review Vocabulary**

frequency: number of wavelengths that pass a given point in one second, measured in hertz (Hz)

New Vocabulary

- loudness
- pitch
- echo
- Doppler effect

Sound and Vibration

Think of all the sounds you've heard since you awoke this morning. Did you hear your alarm clock blaring, car horns honking, or locker doors slamming? Every sound has something in common with every other sound. Each is produced by something that vibrates.

Sound Waves How does an object that is vibrating produce sound? When you speak, the vocal cords in your throat vibrate. These vibrations cause other people to hear your voice. The vibrations produce sound waves that travel to their ears. The other person's ears interpret these sound waves.

A wave carries energy from one place to another without transferring matter. An object that is vibrating in air, such as your vocal cords, produces a sound wave. The vibrating object causes air molecules to move back and forth. As these air molecules collide with those nearby, they cause other air molecules to move back and forth. In this way, energy is transferred from one place to another. A sound wave is a compressional wave, like the wave moving through the coiled spring toy in **Figure 1.** In a compressional wave, particles in the material move back and forth along the direction the wave is moving. In a sound wave, air molecules move back and forth along the direction the sound wave is moving.

Figure 1 When the coils of a coiled spring toy are squeezed together, a compressional wave moves along the spring. The coils move back and forth as the compressional wave moves past them.

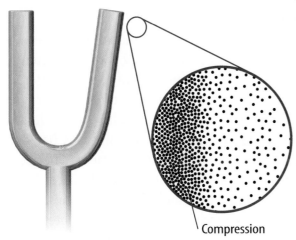

Compression

When the tuning fork vibrates outward, it forces molecules in the air next to it closer together, creating a region of compression.

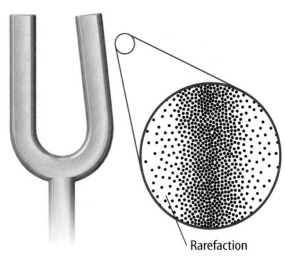

Rarefaction

When the tuning fork moves back, the molecules in the air next to it spread farther apart, creating a region of rarefaction.

Making Sound Waves When an object vibrates, it exerts a force on the surrounding air. For example, as the end of the tuning fork moves outward into the air, it pushes the molecules in the air together, as shown on the left in **Figure 2.** As a result, a region where the molecules are closer together, or more dense, is created. This region of higher density is called a compression. When the end of the tuning fork moves back, it creates a region of lower density called a rarefaction, as shown on the right in **Figure 2.** As the tuning fork continues to vibrate, a series of compressions and rarefactions is formed. The compressions and rarefactions move away from the tuning fork as molecules in these regions collide with other nearby molecules.

Like other waves, a sound wave can be described by its wavelength and frequency. The wavelength of a sound wave is shown in **Figure 3.** The frequency of a sound wave is the number of compressions or rarefactions that pass by a given point in one second. An object that vibrates faster forms a sound wave with a higher frequency.

Figure 2 A tuning fork makes a sound wave as the ends of the fork vibrate in the air.
Explain *why a sound wave cannot travel in a vacuum.*

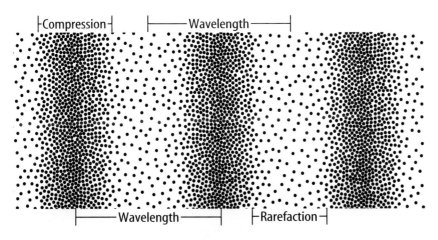

Compression
Wavelength
Wavelength
Rarefaction

Figure 3 Wavelength is the distance from one compression to another or one rarefaction to another.

Mini LAB

Comparing and Contrasting Sounds

Procedure

1. Strike a **block of wood** with a **spoon** and listen carefully to the sound. Then press the block of wood to your ear and strike it with the spoon again. Listen carefully to the sound.
2. Tie the middle of a length of **cotton string** to a metal spoon. Strike the spoon on something to hear it ring. Now press the ends of the string against your ears and repeat the experiment. What do you hear?

Analysis

1. Did you hear sounds transmitted through wood and through string? Describe the sounds.
2. Compare and contrast the sounds in wood and in air.

Try at Home

The Speed of Sound

Sound waves can travel through other materials besides air. In fact, sound waves travel in the same way through different materials as they do in air, although they might travel at different speeds. As a sound wave travels through a material, the particles in the material collide with each other. In a solid, molecules are closer together than in liquids or gases, so collisions between molecules occur more rapidly than in liquids or gases. The speed of sound is usually fastest in solids, where molecules are closest together, and slowest in gases, where molecules are farthest apart. **Table 1** shows the speed of sound through different materials.

The Speed of Sound and Temperature The temperature of the material that sound waves are traveling through also affects the speed of sound. As a substance heats up, its molecules move faster, so they collide more frequently. The more frequent the collisions are, the faster the speed of sound is in the material. For example, the speed of sound in air at 0°C is 331 m/s; at 20°C, it is 343 m/s.

Amplitude and Loudness

What's the difference between loud sounds and quiet sounds? When you play a song at high volume and low volume, you hear the same instruments and voices, but something is different. The difference is that loud sound waves generally carry more energy than soft sound waves do.

Loudness is the human perception of how much energy a sound wave carries. Not all sound waves with the same energy are as loud. Humans hear sounds with frequencies between 3,000 Hz and 4,000 Hz as being louder than other sound waves with the same energy.

Table 1 Speed of Sound Through Different Materials	
Material	**Speed (m/s)**
Air	343
Water	1,483
Steel	5,940
Glass	5,640

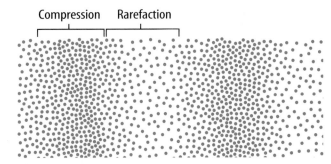

Compression   Rarefaction

This sound wave has a lower amplitude.

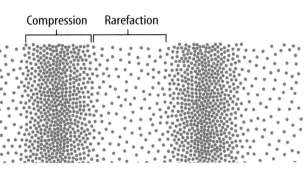

Compression   Rarefaction

This sound wave has a higher amplitude. Particles in the material are more compressed in the compressions and more spread out in the rarefactions.

Amplitude and Energy The amount of energy a wave carries depends on its amplitude. For a compressional wave such as a sound wave, the amplitude is related to how spread out the molecules or particles are in the compressions and rarefactions, as **Figure 4** shows. The higher the amplitude of the wave is, the more compressed the particles in the compression are and the more spread out they are in the rarefactions. More energy had to be transferred by the vibrating object that created the wave to force the particles closer together or spread them farther apart. Sound waves with greater amplitude carry more energy and sound louder. Sound waves with smaller amplitude carry less energy and sound quieter.

Figure 4 The amplitude of a sound wave depends on how spread out the particles are in the compressions and rarefactions of the wave.

Reading Check *What determines the loudness of different sounds?*

Figure 5 The loudness of sound is measured on the decibel scale.

The Decibel Scale Perhaps an adult has said to you, "Turn down your music, it's too loud! You're going to lose your hearing!" Although the perception of loudness varies from person to person, the energy carried by sound waves can be described by a scale called the decibel (dB) scale. **Figure 5** shows the decibel scale. An increase in the loudness of a sound of 10 dB means that the energy carried by the sound has increased ten times, but an increase of 20 dB means that the sound carries 100 times more energy.

Hearing damage begins to occur at sound levels of about 85 dB. The amount of damage depends on the frequencies of the sound and the length of time a person is exposed to the sound. Some music concerts produce sound levels as high as 120 dB. The energy carried by these sound waves is about 30 billion times greater than the energy carried by sound waves that are made by whispering.

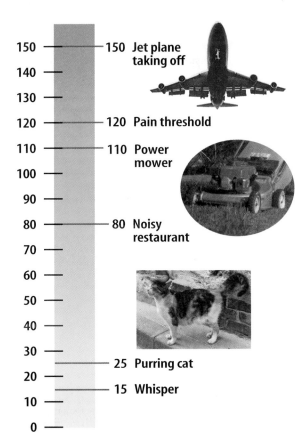

150	150 Jet plane taking off
140	
130	
120	120 Pain threshold
110	110 Power mower
100	
90	
80	80 Noisy restaurant
70	
60	
50	
40	
30	
	25 Purring cat
20	
	15 Whisper
10	
0	

Frequency and Pitch

The **pitch** of a sound is how high or low it sounds. For example, a piccolo produces a high-pitched sound or tone, and a tuba makes a low-pitched sound. Pitch corresponds to the frequency of the sound. The higher the pitch is, the higher the frequency is. A sound wave with a frequency of 440 Hz, for example, has a higher pitch than a sound wave with a frequency of 220 Hz.

The human ear can detect sound waves with frequencies between about 20 Hz and 20,000 Hz. However, some animals can detect even higher and lower frequencies. For example, dogs can hear frequencies up to almost 50,000 Hz. Dolphins and bats can hear frequencies as high as 150,000 Hz, and whales can hear frequencies higher than those heard by humans.

Recall that frequency and wavelength are related. If two sound waves are traveling at the same speed, the wave with the shorter wavelength has a higher frequency. If the wavelength is shorter, then more compressions and rarefactions will go past a given point every second than for a wave with a longer wavelength, as shown in **Figure 6.** Sound waves with a higher pitch have shorter wavelengths than those with a lower pitch.

The Human Voice When you make a sound, you exhale past your vocal cords, causing them to vibrate. The length and thickness of your vocal cords help determine the pitch of your voice. Shorter, thinner vocal cords vibrate at higher frequencies than longer or thicker ones. This explains why children, whose vocal cords are still growing, have higher voices than adults. Muscles in the throat can stretch the vocal cords tighter, letting people vary their pitch within a limited range.

Figure 6 The upper sound wave has a shorter wavelength than the lower wave. If these two sound waves are traveling at the same speed, the upper sound wave has a higher frequency than the lower one. For this wave, more compressions and rarefactions will go past a point every second than for the lower wave.
Identify *the wave that has a higher pitch.*

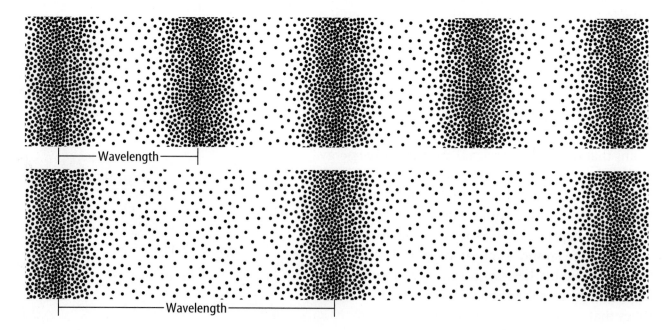

Wavelength

Wavelength

Figure 7 Sonar uses reflected sound waves to determine the location and shape of an object.

Echoes

Sound reflects off of hard surfaces, just like a water wave bounces off the side of a bath tub. A reflected sound wave is called an **echo.** If the distance between you and a reflecting surface is great enough, you might hear the echo of your voice. This is because it might take a few seconds for the sound to travel to the reflecting surface and back to your ears.

Sonar systems use sound waves to map objects underwater, as shown in **Figure 7.** The amount of time it takes an echo to return depends on how far away the reflecting surface is. By measuring the length of time between emitting a pulse of sound and hearing its echo off the ocean floor, the distance to the ocean floor can be measured. Using this method, sonar can map the ocean floor and other undersea features. Sonar also can be used to detect submarines, schools of fish, and other objects.

 **Reading Check** *How do sonar systems measure distance?*

INTEGRATE Life Science  **Echolocation** Some animals use a method called echolocation to navigate and hunt. Bats, for example, emit high-pitched squeaks and listen for the echoes. The type of echo it hears helps the bat determine exactly where an insect is, as shown in **Figure 8.** Dolphins also use a form of echolocation. Their high-pitched clicks bounce off of objects in the ocean, allowing them to navigate in the same way.

People with visual impairments also use echolocation. For example, they can interpret echoes to estimate the size and shape of a room by using their ears.

Science Online

Topic: Sonar
Visit booko.msscience.com for Web links to information about how sonar is used to detect objects underwater.

Activity List and explain how several underwater discoveries were made using sonar.

Figure 8 Bats use echolocation to hunt.
Explain *why this is a good technique for hunting at night.*

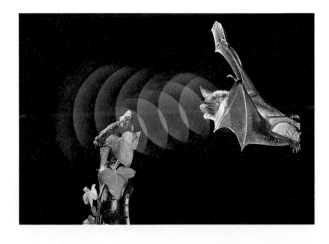

The Doppler Effect

Perhaps you've heard an ambulance siren as the ambulance speeds toward you, then goes past. You might have noticed that the pitch of the siren gets higher as the ambulance moves toward you. Then as the ambulance moves away, the pitch of the siren gets lower. The change in frequency that occurs when a source of sound is moving relative to a listener is called the **Doppler effect. Figure 9** shows why the Doppler effect occurs.

The Doppler effect occurs whether the sound source or the listener is moving. If you drive past a factory as its whistle blows, the whistle will sound higher pitched as you approach. As you move closer you encounter each sound wave a little earlier than you would if you were sitting still, so the whistle has a higher pitch. When you move away from the whistle, each sound wave takes a little longer to reach you. You hear fewer wavelengths per second, which makes the sound lower in pitch.

Radar guns that are used to measure the speed of cars and baseball pitches also use the Doppler effect. Instead of a sound wave, the radar gun sends out a radio wave. When the radio wave is reflected, its frequency changes depending on the speed of the object and whether it is moving toward the gun or away from it. The radar gun uses the change in frequency of the reflected wave to determine the object's speed.

Doppler Shift of Light
The frequency of light waves is also changed by the Doppler shift. If a light source is moving away from an observer, the frequencies of the emitted light waves decrease. Research how the Doppler shift is used by astronomers to determine how other objects in the universe are moving relative to Earth.

Applying Science

How does Doppler radar work?

Doppler radar is used by the National Weather Service to detect areas of precipitation and to measure the speed at which a storm moves. Because the wind moves the rain, Doppler radar can "see" into a strong storm and expose the winds. Tornadoes that might be forming in the storm then can be identified.

Identify the Problem

An antenna sends out pulses of radio waves as it rotates. The waves bounce off raindrops and return to the antenna at a different frequency, depending on whether the rain is moving toward the antenna or away from it. The change in frequency is due to the Doppler shift.

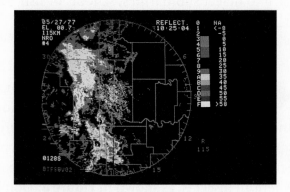

Solving the Problem

1. If the frequency of the reflected radio waves increases, how is the rain moving relative to the radar station?
2. In a tornado, winds are rotating. How would the radio waves reflected by rotating winds be Doppler-shifted?

Figure 9

Y ou've probably heard the siren of an ambulance as it races through the streets. The sound of the siren seems to be higher in pitch as the ambulance approaches and lower in pitch as it moves away. This is the Doppler effect, which occurs when a listener and a source of sound waves are moving relative to each other.

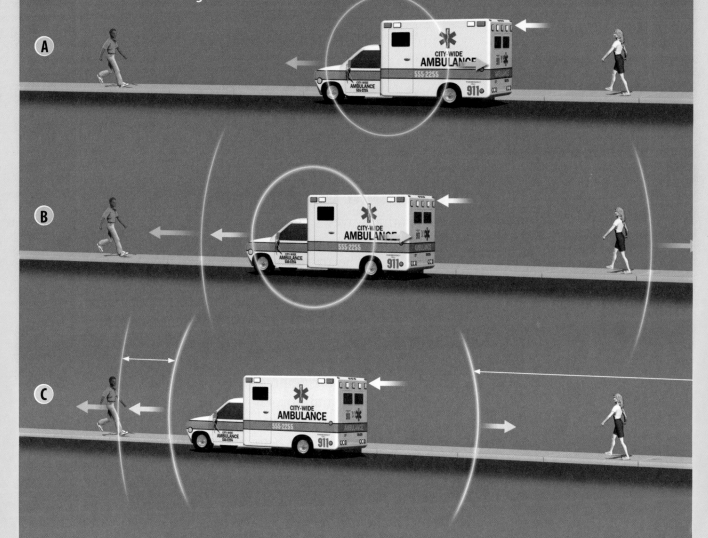

A As the ambulance speeds down the street, its siren emits sound waves. Suppose the siren emits the compression part of a sound wave as it goes past the girl.

B As the ambulance continues moving, it emits another compression. Meanwhile, the first compression spreads out from the point from which it was emitted.

C The waves traveling in the direction that the ambulance is moving have compressions closer together. As a result, the wavelength is shorter and the boy hears a higher frequency sound as the ambulance moves toward him. The waves traveling in the opposite direction have compressions that are farther apart. The wavelength is longer and the girl hears a lower frequency sound as the ambulance moves away from her.

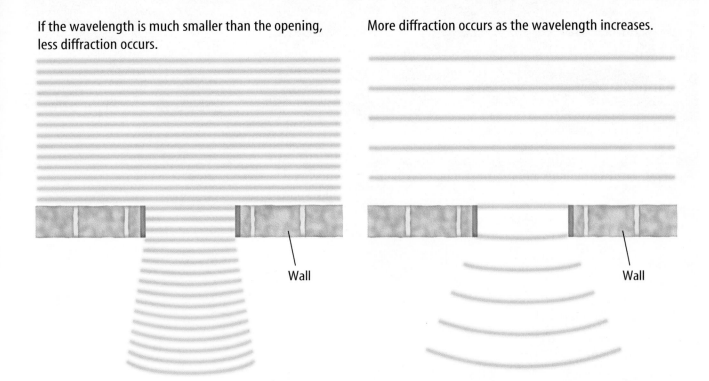

If the wavelength is much smaller than the opening, less diffraction occurs.

More diffraction occurs as the wavelength increases.

Wall

Wall

Figure 10 The spreading of a wave by diffraction depends on the wavelength and the size of the opening.

Diffraction of Sound Waves

Like other waves, sound waves diffract. This means they can bend around obstacles or spread out after passing through narrow openings. The amount of diffraction depends on the wavelength of the sound wave compared to the size of the obstacle or opening. If the wavelength is much smaller than the obstacle, almost no diffraction occurs. As the wavelength becomes closer to the size of the obstacle, the amount of diffraction increases.

You can observe diffraction of sound waves by visiting the school band room during practice. If you stand in the doorway, you will hear the band normally. However, if you stand to one side outside the door or around a corner, you will hear the lower-pitched instruments better. **Figure 10** shows why this happens. The sound waves that are produced by the lower-pitched instruments have lower frequencies and longer wavelengths. These wavelengths are closer to the size of the door opening than the higher-pitched sound waves are. As a result, the longer wavelengths diffract more, and you can hear them even when you're not standing in the doorway.

The diffraction of lower frequencies in the human voice allows you to hear someone talking even when the person is around the corner. This is different from an echo. Echoes occur when sound waves bounce off a reflecting surface. Diffraction occurs when a wave spreads out after passing through an opening, or when a wave bends around an obstacle.

Using Sound Waves

Sound waves can be used to treat certain medical problems. A process called ultrasound uses high-frequency sound waves as an alternative to some surgeries. For example, some people develop small, hard deposits in their kidneys or gallbladders. A doctor can focus ultrasound waves at the kidney or gallbladder. The ultrasound waves cause the deposits to vibrate rapidly until they break apart into small pieces. Then, the body can get rid of them.

Ultrasound can be used to make images of the inside of the body. One common use of ultrasound is to examine a developing fetus. Also, ultrasound along with the Doppler effect can be used to examine the functioning of the heart. An ultrasound image of the heart is shown in **Figure 11.** This technique can help determine if the heart valves and heart muscle are functioning properly, and how blood is flowing through the heart.

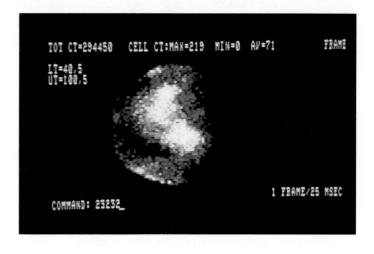

Figure 11 Ultrasound is used to make this image of the heart. **Describe** *other ways ultrasound is used in medicine.*

section 1 review

Summary

Sound Waves

- Sound waves are compressional waves produced by vibrations.
- Sound travels fastest in solids and slowest in gases.
- Sound travels faster as the temperature of the medium increases.
- The energy carried by a sound wave increases as its amplitude increases.

Loudness and Pitch

- Loudness is the human perception of the energy carried by a sound wave.
- The pitch of a sound becomes higher as the frequency of the sound increases.

The Doppler Effect and Diffraction

- In the Doppler effect, the frequency of a sound wave changes if the source of the sound is moving relative to the listener.
- Diffraction occurs when sound waves bend around objects or spread out after passing through an opening.

Self Check

1. **Describe** how the loudness of a sound wave changes when the amplitude of the wave is increased.
2. **Explain** how the wavelength of a sound wave affects the diffraction of the sound wave through an open window.
3. **Describe** how echolocation could be used to measure the distance to the bottom of a lake.
4. **Discuss** how the spacing of particles in a sound wave changes as the amplitude of the wave decreases.
5. **Describe** how the wavelength of a sound wave changes if the frequency of the wave increases.
6. **Think Critically** You hear the pitch of the sound from an ambulance siren get lower, then get higher. Describe the motion of the ambulance relative to you.

Applying Math

7. **Calculate Distance** Sound travels through water at a speed of 1,483 m/s. Use the equation

   distance = speed × time

   to calculate how far a sound wave in water will travel in 5 s.

Observe and Measure Reflection of Sound

▶ Real-World Question

Like all waves, sound waves can be reflected. When sound waves strike a surface, in what direction does the reflected sound wave travel? In this activity, you'll focus sound waves using cardboard tubes to help answer this question. How are the angles made by incoming and reflected sound waves related?

Goals

- ■ **Observe** reflection of sound waves.
- ■ **Measure** the angles incoming and reflected sound waves make with a surface.

Materials

20-cm to 30-cm-long cardboard tubes (2)
watch that ticks audibly
protractor

Safety Precautions

▶ Procedure

1. Work in groups of three. Each person should listen to the watch—first without a tube and then through a tube. The person who hears the watch most easily is the listener.

2. One person should hold one tube at an angle with one end above a table. Hold the watch at the other end of the tube.

3. The listener should hold the second tube at an angle, with one end near his or her ear and the other end near the end of the first tube that is just above the table. The tubes should be in the same vertical plane.

4. Move the first tube until the watch sounds

loudest. The listener might need to cover the other ear to block out background noises.

5. The third person should measure the angle that each tube makes with the table.

▶ Conclude and Apply

1. **Compare** the angles the incoming and reflected waves make with the table.

2. The normal is a line at 90 degrees to the table at the point where reflection occurs. Determine the angles the incoming and reflected waves make with the normal.

3. The law of reflection states that the angles the incoming and reflected waves make with the normal are equal. Do sound waves obey the law of reflection?

𝒞ommunicating Your Data

Make a scientific illustration to show how the experiment was done. Describe your results using the illustration.

Music

What is music?

What do you like to listen to—rock 'n' roll, country, blues, jazz, rap, or classical? Music and noise are groups of sounds. Why do humans hear some sounds as music and other sounds as noise?

The answer involves patterns of sound. **Music** is a group of sounds that have been deliberately produced to make a regular pattern. Look at **Figure 12.** The sounds that make up music usually have a regular pattern of pitches, or notes. Some natural sounds such as the patter of rain on a roof, the sound of ocean waves splashing, or the songs of birds can sound musical. On the other hand, noise is usually a group of sounds with no regular pattern. Sounds you hear as noise are irregular and disorganized such as the sounds of traffic on a city street or the roar of a jet aircraft.

However, the difference between music and noise can vary from person to person. What one person considers to be music, another person might consider noise.

Natural Frequencies Music is created by vibrations. When you sing, your vocal cords vibrate. When you beat a drum, the drumhead vibrates. When you play a guitar, the strings vibrate.

If you tap on a bell with a hard object, the bell produces a sound. When you tap on a bell that is larger or smaller or has a different shape you hear a different sound. The bells sound different because each bell vibrates at different frequencies. A bell vibrates at frequencies that depend on its shape and the material it is made from. Every object will vibrate at certain frequencies called its **natural frequencies.**

as you read

What You'll Learn

- **Explain** the difference between music and noise.
- **Describe** how different instruments produce music.
- **Explain** how you hear.

Why It's Important

Music is made by people in every part of the world.

🔎 Review Vocabulary

compressional wave: a type of mechanical wave in which matter in the medium moves forward and backward in the same direction the wave travels

New Vocabulary

- music
- natural frequencies
- resonance
- fundamental frequency
- overtone
- reverberation
- eardrum

Figure 12 Music and noise have different types of sound patterns.

Noise has no specific or regular sound wave pattern.

Music is organized sound. Music has regular sound wave patterns and structures.

Musical Instruments and Natural Frequencies Many objects vibrate at one or more natural frequencies when they are struck or disturbed. Like a bell, the natural frequencies of any object depend on the size and shape of the object and the material it is made from. Musical instruments use the natural frequencies of strings, drumheads, or columns of air contained in pipes to produce various musical notes.

✓ **Reading Check** *What determines the natural frequencies?*

Resonance You may have seen the comedy routine in which a loud soprano sings high enough to shatter glass. Sometimes sound waves cause an object to vibrate. When a tuning fork is struck, it vibrates at its natural frequency and produces a sound wave with the same frequency. Suppose you have two tuning forks with the same natural frequency. You strike one tuning fork, and the sound waves it produces strike the other tuning fork. These sound waves would cause the tuning fork that wasn't struck to absorb energy and vibrate. This is an example of resonance. **Resonance** occurs when an object is made to vibrate at its natural frequencies by absorbing energy from a sound wave or another object vibrating at these frequencies.

Musical instruments use resonance to amplify their sounds. Look at **Figure 13.** The vibrating tuning fork might cause the table to vibrate at the same frequency, or resonate. The combined vibrations of the table and the tuning fork increase the loudness of the sound waves produced.

Reducing Earthquake Damage The shaking of the ground during an earthquake can cause buildings to resonate. The increased vibration of a building due to resonance could result in the collapse of the building, causing injuries and loss of life. To reduce damage during earthquakes, buildings are designed to resonate at frequencies different than those that occur during earthquakes. Research how buildings are designed to reduce damage caused by earthquakes.

Figure 13 When a vibrating tuning fork is placed against a table, resonance might cause the table to vibrate.

Overtones

Before a concert, all orchestra musicians tune their instruments by playing the same note. Even though the note has the same pitch, it sounds different for each instrument. It also sounds different from a tuning fork that vibrates at the same frequency as the note.

A tuning fork produces a single frequency, called a pure tone. However, the notes produced by musical instruments are not pure tones. Most objects have more than one natural frequency at which they can vibrate. As a result, they produce sound waves of more than one frequency.

If you play a single note on a guitar, the pitch that you hear is the lowest frequency produced by the vibrating string. The lowest frequency produced by a vibrating object is the **fundamental frequency.** The vibrating string also produces higher frequencies. These higher frequencies are **overtones.** Overtones have frequencies that are multiples of the fundamental frequency, as in **Figure 14.** The number and intensity of the overtones produced by each instrument are different and give instruments their distinctive sound quality.

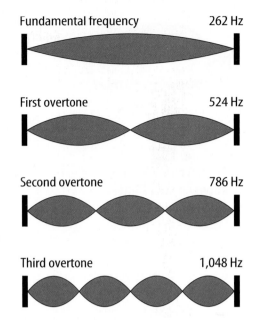

Fundamental frequency	262 Hz
First overtone	524 Hz
Second overtone	786 Hz
Third overtone	1,048 Hz

Figure 14 A string vibrates at a fundamental frequency, as well as at overtones. The overtones are multiples of that frequency.

Musical Scales

A musical instrument is a device that produces musical sounds. These sounds are usually part of a musical scale that is a sequence of notes with certain frequencies. For example, **Figure 15** shows the sequence of notes that belong to the musical scale of C. Notice that the frequency produced by the instrument doubles after eight successive notes of the scale are played. Other musical scales consist of a different sequence of frequencies.

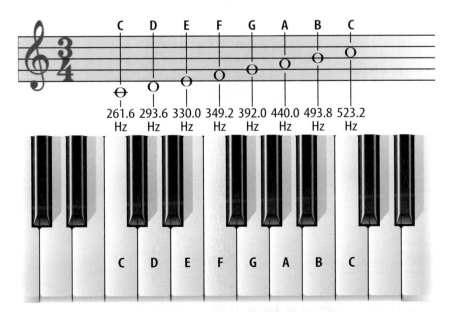

Figure 15 A piano produces a sequence of notes that are a part of a musical scale.
Describe how the frequencies of the two C notes on this scale are related.

Modeling a Stringed Instrument

Procedure 👓

1. Stretch a **rubber band** between your fingers.
2. Pluck the rubber band. Listen to the sound and observe the shape of the vibrating band. Record what you hear and see.
3. Stretch the band farther and repeat step 2.
4. Shorten the length of the band that can vibrate by holding your finger on one point. Repeat step 2.
5. Stretch the rubber band over an open box, such as a **shoe box**. Repeat step 2.

Analysis

1. How did the sound change when you stretched the rubber band? Was this what you expected? Explain.
2. How did the sound change when you stretched the band over the box? Did you expect this? Explain.

Stringed Instruments

Stringed instruments, like the cello shown in **Figure 16,** produce music by making strings vibrate. Different methods are used to make the strings vibrate—guitar strings are plucked, piano strings are struck, and a bow is slid across cello strings. The strings often are made of wire. The pitch of the note depends on the length, diameter, and tension of the string—if the string is shorter, narrower, or tighter, the pitch increases. For example, pressing down on a vibrating guitar string shortens its length and produces a note with a higher pitch. Similarly, the thinner guitar strings produce a higher pitch than the thicker strings.

Amplifying Vibrations The sound produced by a vibrating string usually is soft. To amplify the sound, stringed instruments usually have a hollow chamber, or box, called a resonator, which contains air. The resonator absorbs energy from the vibrating string and vibrates at its natural frequencies. For example, the body of a guitar is a resonator that amplifies the sound that is produced by the vibrating strings. The vibrating strings cause the guitar's body and the air inside it to resonate. As a result, the vibrating guitar strings sound louder, just as the tuning fork that was placed against the table sounded louder.

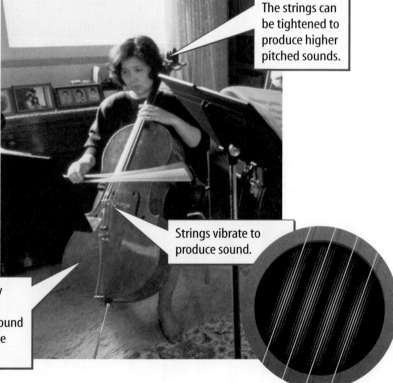

Figure 16 A cello is a stringed instrument. When strings vibrate, the natural frequencies of the instrument's body amplify the sound.

The strings can be tightened to produce higher pitched sounds.

Strings vibrate to produce sound.

The cello's body resonates and amplifies the sound produced by the strings.

Percussion

Percussion instruments, such as the drum shown in **Figure 17,** are struck to make a sound. Striking the top surface of the drum causes it to vibrate. The vibrating drumhead is attached to a chamber that resonates and amplifies the sound.

Drums and Pitch Some drums have a fixed pitch, but some can be tuned to play different notes. For example, if the drumhead on a kettledrum is tightened, the natural frequency of the drumhead is increased. As a result, the pitches of the sounds produced by the kettledrum get higher. A steel drum, shown in **Figure 17,** plays different notes in the scale when different areas in the drum are struck. In a xylophone, wood or metal bars of different lengths are struck. The longer the bar is, the lower the note that it produces is.

Figure 17 The sounds produced by drums depend on the material that is vibrating.

The vibrating drumhead of this drum is amplified by the resonating air in the body of the drum.

The vibrating steel surface in a steel drum produces loud sounds that don't need to be amplified by an air-filled chamber.

Brass and Woodwinds

Just as the bars of a xylophone have different natural frequencies, so do the air columns in pipes of different lengths. Brass and woodwind instruments, such as those in **Figure 18,** are essentially pipes or tubes of different lengths that sometimes are twisted around to make them easier to hold and carry. To make music from these instruments, the air in the pipes is made to vibrate at various frequencies.

Different methods are used to make the air column vibrate. A musician playing a brass instrument, such as a trumpet, makes the air column vibrate by vibrating the lips and blowing into the mouthpiece. Woodwinds such as clarinets, saxophones, and oboes contain one or two reeds in the mouthpiece that vibrate the air column when the musician blows into the mouthpiece. Flutes also are woodwinds, but a flute player blows across a narrow opening to make the air column vibrate.

Figure 18 Brass and woodwind instruments produce sounds by causing a column of air to vibrate.

Figure 19 A flute changes pitch as holes are opened and closed.

By opening holes on a flute, the length of the vibrating air column is made shorter.

Changing Pitch in Woodwinds To change the note that is being played in a woodwind instrument, a musician changes the length of the resonating column of air. By making the length of the vibrating air column shorter, the pitch of the sound produced is made higher. In a woodwind such as a flute, saxophone, or clarinet, this is done by closing and opening finger holes along the length of the instrument, as shown in **Figure 19.**

Changing Pitch in Brass In brass instruments, musicians vary the pitch in other ways. One is by blowing harder to make the air resonate at a higher natural frequency. Another way is by pressing valves that change the length of the tube.

Beats

Recall that interference occurs when two waves overlap and combine to form a new wave. The new wave formed by interference can have a different frequency, wavelength, and amplitude than the two original waves.

Suppose two notes close in frequency are played at the same time. The two notes interfere to form a new sound whose loudness increases and decreases several times a second. If you were listening to the sound, you would hear a series of beats as the sound got louder and softer. The beat frequency, or the number of beats you would hear each second, is equal to the difference in the frequencies of the two notes.

For example, if the two notes have frequencies of 329 Hz and 332 Hz, the beat frequency would be 3 Hz. You would hear the sound get louder and softer—a beat—three times each second.

Figure 20 A piano can be tuned by using beats.

Beats Help Tune Instruments Beats are used to help tune instruments. For example, a piano tuner, like the one shown in **Figure 20,** might hit a tuning fork and then the corresponding key on the piano. Beats are heard when the difference in pitch is small. The piano string is tuned properly when the beats disappear. You might have heard beats while listening to an orchestra tune before a performance. You also can hear beats produced by two engines vibrating at slightly different frequencies.

Reverberation

Sound is reflected by hard surfaces. In an empty gymnasium, the sound of your voice can be reflected back and forth several times by the floor, walls, and ceiling. Repeated echoes of sound are called **reverberation.** In a gym, reverberation makes the sound of your voice linger before it dies out. Some reverberation can make voices or music sound bright and lively. Too little reverberation makes the sound flat and lifeless. However, reverberation can produce a confusing mess of noise if too many sounds linger for too long.

Concert halls and theaters, such as the one in **Figure 21,** are designed to produce the appropriate level of reverberation. Acoustical engineers use soft materials to reduce echoes. Special panels that are attached to the walls or suspended from the ceiling are designed to reflect sound toward the audience.

Science Online

Topic: Controlling Reverberation
Visit booko.msscience.com for Web Links to information about how acoustical engineers control reverberation.

Activity Make a list of the materials engineers use to reduce and enhance reverberation.

Figure 21 The shape of a concert hall and the materials it contains are designed to control the reflection of sound waves.

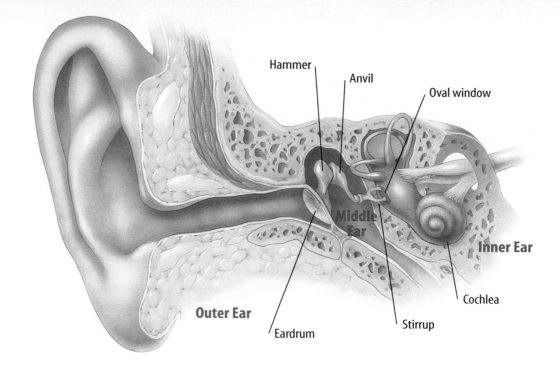

Hammer

Anvil

Oval window

Middle Ear

Inner Ear

Outer Ear

Cochlea

Stirrup

Eardrum

Figure 22 The human ear has three different parts—the outer ear, the middle ear, and the inner ear.

Figure 23 Animals, such as rabbits and owls, have ears that are adapted to their different needs.

The Ear

You hear sounds with your ears. The ear is a complex organ that is able to detect a wide range of sounds. The ear can detect frequencies ranging from about 20 Hz to about 20,000 Hz. The ear also can detect a wide range of sound intensities. The faintest sounds you can hear carry about one trillionth the amount of energy as the loudest sounds you can hear. The human ear is illustrated in **Figure 22.** It has three parts—the outer ear, the middle ear, and the inner ear.

The Outer Ear—Sound Collector Your outer ear collects sound waves and directs them into the ear canal. Notice that your outer ear is shaped roughly like a funnel. This shape helps collect sound waves.

Animals that rely on hearing to locate predators or prey often have larger, more adjustable ears than humans, as shown in **Figure 23.** A barn owl, which relies on its excellent hearing for hunting at night, does not have outer ears made of flesh. Instead, the arrangement of its facial feathers helps direct sound to its ears. Some sea mammals, on the other hand, have only small holes for outer ears, even though their hearing is good.

The Middle Ear—Sound Amplifier When sound waves reach the middle ear, they vibrate the **eardrum,** which is a membrane that stretches across the ear canal like a drumhead. When the eardrum vibrates, it transmits vibrations to three small connected bones—the hammer, anvil, and stirrup. The bones amplify the vibrations, just as a lever can change a small movement at one end into a larger movement at the other.

The Inner Ear—Sound Interpreter The stirrup vibrates a second membrane called the oval window. This marks the start of the inner ear, which is filled with fluid. Vibrations in the fluid are transmitted to hair-tipped cells lining the cochlea, as shown in **Figure 24.** Different sounds vibrate the cells in different ways. The cells generate signals containing information about the frequency, intensity, and duration of the sound. The nerve impulses travel along the auditory nerve and are transmitted to the part of the brain that is responsible for hearing.

✓ Reading Check *Where are waves detected and interpreted in the ear?*

Figure 24 The inner ear contains tiny hair cells that convert vibrations into nerve impulses that travel to the brain.

Hearing Loss

The ear can be damaged by disease, age, and exposure to loud sounds. For example, constant exposure to loud noise can damage hair cells in the cochlea. If damaged mammalian hair cells die, some loss of hearing results because mammals cannot make new hair cells. Also, some hair cells and nerve fibers in the inner ear degenerate and are lost as people age. It is estimated that about 30 percent of people over 65 have some hearing loss due to aging.

section 2 review

Summary

What is music?
- Music is sound that is deliberately produced in a regular pattern.
- Objects vibrate at certain natural frequencies.
- The lowest frequency produced by a vibrating object is the object's fundamental frequency.
- The overtones produced by a vibrating object are multiples of the fundamental frequency.

Musical Instruments and Hearing
- In stringed instruments the sounds made by vibrating strings are amplified by a resonator.
- Percussion instruments produce sound by vibrating when they are struck.
- Brass and woodwind instruments produce sound by vibrating a column of air.
- The ear collects sound waves, amplifies the sound, and interprets the sound.

Self Check

1. **Describe** how music and noise are different.
2. **Infer** Two bars on a xylophone are 10 cm long and 14 cm long. Identify which bar produces a lower pitch when struck and explain why.
3. **Describe** the parts of the human ear and the function of each part in enabling you to hear sound.
4. **Predict** how the sound produced by a guitar string changes as the length of the string is made shorter.
5. **Diagram** the fundamental and the first two overtones for a vibrating string.
6. **Think Critically** How does reverberation explain why your voice sounds different in a gym than it does in your living room?

Applying Math

7. **Calculate Overtone Frequency** A guitar string has a fundamental frequency of 440 Hz. What is the frequency of the second overtone?

Music

Goals

- **Design** an experiment to compare the changes that are needed in different instruments to produce a variety of different notes.
- **Observe** which changes are made when playing different notes.
- **Measure and record** these changes whenever possible.

Possible Materials

musical instruments
measuring tape
tuning forks

Safety Precautions

Properly clean the mouthpiece of any instrument before it is used by another student.

⊙ Real-World Question

The pitch of a note that is played on an instrument sometimes depends on the length of the string, the air column, or some other vibrating part. Exactly how does sound correspond to the size or length of the vibrating part? Is this true for different instruments? What causes different instruments to produce different notes?

⊙ Form a Hypothesis

Based on your reading and observations, make a hypothesis about what changes in an instrument to produce different notes.

⊙ Test Your Hypothesis

Make a Plan

1. You should do this lab as a class, using as many instruments as possible. You might want to go to the music room or invite friends and relatives who play an instrument to visit the class.

2. As a group, decide how you will measure changes in instruments. For wind instruments, can you measure the length of the vibrating air column? For stringed instruments, can you measure the length and thickness of the vibrating string?

3. Refer to the table of wavelengths and frequencies for notes in the scale. Note that no measurements are given—if you measure C to correspond to a string length of 30 cm, for example, the note G will correspond to two thirds of that length.

4. Decide which musical notes you will compare. Prepare a table to collect your data. List the notes you have selected.

Ratios of Wavelengths and Frequencies of Musical Notes		
Note	Wavelength	Frequency
C	1	1
D	8/9	9/8
E	4/5	5/4
F	3/4	4/3
G	2/3	3/2
A	3/5	5/3
B	8/15	15/8
C	1/2	2

Follow Your Plan

1. Make sure your teacher approves your plan before you start.

2. Carry out the experiment as planned.

3. While doing the experiment, record your observations and complete the data table.

Analyze Your Data

1. **Compare** the change in each instrument when the two notes are produced.

2. **Compare and contrast** the changes between instruments.

3. What were the controls in this experiment?

4. What were the variables in this experiment?

5. How did you eliminate bias?

Conclude and Apply

1. How does changing the length of the vibrating column of air in a wind instrument affect the note that is played?

2. Describe how you would modify an instrument to increase the pitch of a note that is played.

Communicating Your Data

Demonstrate to another teacher or to family members how the change in the instrument produces a change in sound.

It's a Wrap!

No matter how quickly or slowly you open a candy wrapper, it always will make a noise

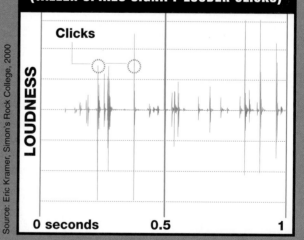

crackle! snap! snap! snap! crackle! pop! pop! pop! crackle!

Y ou're at the movies and it's the most exciting part of the film. The audience is silent and intent on what is happening on the screen. At that moment, you decide to unwrap a piece of candy. CRACKLE! POP! SNAP! No matter how you do it, the candy wrapper makes a lot of noise.

Why can't you unwrap candy without making a racket? To test this plastics problem, researchers put some crinkly wrappers in a silent room. Then they stretched out the wrappers and recorded the sounds they made. Next, the sounds were analyzed by a computer. The research team discovered that the wrapper didn't make a continuous sound. Instead, it made many separate little popping noises, each taking only a

thousandth of a second. They found that whether you open the wrapper quickly or slowly the amount of noise made by the pops will be the same. "And there's nothing you can do about it," said a member of the research team.

By understanding what makes a plastic wrapper snap when it changes shape, doctors can better understand molecules in the human body that also change shape.

The pop chart

SOUND LEVEL OVER TIME

The sound that a candy wrapper makes is emitted as a series of pulses or clicks. So, opening a wrapper slowly only increases the length of time in between clicks, but the amount of noise remains the same.

(TALLER SPIKES SIGNIFY LOUDER CLICKS)

Clicks

LOUDNESS

0 seconds   0.5   1

Source: Eric Kramer, Simon's Rock College, 2000

Recall and Retell Have you ever opened a candy wrapper in a quiet place? Did it bother other people? If so, did you try to open it more slowly? What happened?

Reviewing Main Ideas

Section 1 What is sound?

1. Sound is a compressional wave that travels through matter, such as air. Sound is produced by something that vibrates.

2. The speed of sound depends on the material in which it is traveling.

3. The larger the amplitude of a sound wave, the more energy it carries and the louder the sound.

4. The pitch of a sound wave becomes higher as its frequency increases. Sound waves can reflect and diffract.

5. The Doppler effect occurs when a source of sound and a listener are in motion relative to each other. The pitch of the sound heard by the listener changes.

Section 2 Music

1. Music is made of sounds that are used in a regular pattern. Noise is made of sounds that are irregular and disorganized.

2. Objects vibrate at their natural frequencies. These depend on the shape of the object and the material it's made of.

3. Resonance occurs when an object is made to vibrate by absorbing energy at one of its natural frequencies.

4. Musical instruments produce notes by vibrating at their natural frequencies.

5. Beats occur when two waves of nearly the same frequency interfere.

6. The ear collects sound waves and converts sound waves to nerve impulses.

Visualizing Main Ideas

Copy and complete the following concept map on sound.

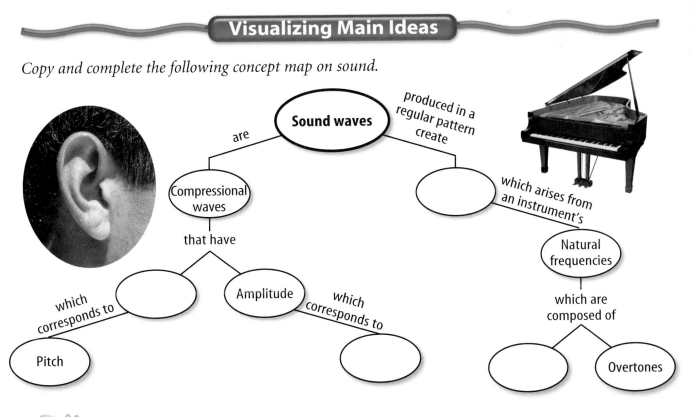

Using Vocabulary

Doppler effect p. 42
eardrum p. 54
echo p. 41
fundamental
  frequency p. 49
loudness p. 38

music p. 47
natural frequency p. 47
overtone p. 49
pitch p. 40
resonance p. 48
reverberation p. 53

Distinguish between the terms in the following pairs.

1. overtones—fundamental frequency

2. pitch—sound wave

3. pitch—Doppler effect

4. loudness—resonance

5. fundamental frequency— natural frequency

6. loudness—amplitude

7. natural frequency—overtone

8. reverberation—resonance

Checking Concepts

Choose the word or phrase that best answers the question.

9. A tone that is lower in pitch is lower in what characteristic?
   A) frequency
   B) wavelength
   C) loudness
   D) resonance

10. If the wave speed stays the same, which of the following decreases as the frequency increases?
   A) pitch
   B) wavelength
   C) loudness
   D) resonance

11. What part of the ear is damaged most easily by continued exposure to loud noise?
   A) eardrum
   B) stirrup
   C) oval window
   D) hair cells

12. What is an echo?
   A) diffracted sound
   B) resonating sound
   C) reflected sound
   D) an overtone

13. A trumpeter depresses keys to make the column of air resonating in the trumpet shorter. What happens to the note being played?
   A) Its pitch is higher.
   B) Its pitch is lower.
   C) It is quieter.
   D) It is louder.

14. When tuning a violin, a string is tightened. What happens to a note being played on the string?
   A) Its pitch is higher.
   B) Its pitch is lower.
   C) It is quieter.
   D) It is louder.

15. As air becomes warmer, how does the speed of sound in air change?
   A) It increases.
   B) It decreases.
   C) It doesn't change.
   D) It oscillates.

16. Sound waves are which type of wave?
   A) slow
   B) transverse
   C) compressional
   D) electromagnetic

17. What does the middle ear do?
   A) focuses sound
   B) interprets sound
   C) collects sound
   D) transmits and amplifies sound

18. An ambulance siren speeds away from you. What happens to the pitch of the siren?
   A) It becomes softer.
   B) It becomes louder.
   C) It decreases.
   D) It increases.

Science Online booko.msscience.com/vocabulary_puzzlemaker

Thinking Critically

19. Explain Some xylophones have open pipes of different lengths hung under each bar. The longer a bar is, the longer the pipe beneath it. Explain how these pipes help amplify the sound of the xylophone.

20. Infer why you don't notice the Doppler effect for a slow moving train.

21. Predict Suppose the movement of the bones in the middle ear were reduced. Which would be more affected—the ability to hear quiet sounds or the ability to hear high frequencies? Explain your answer.

22. Explain The triangle is a percussion instrument consisting of an open metal triangle hanging from a string. A chiming sound is heard when the triangle is struck by a metal rod. If the triangle is held in the hand, a quiet dull sound is heard when it is struck. Why does holding the triangle make the sound quieter?

Use the table below to answer question 23.

Speed of Sound Through Different Materials	
Material	**Speed (m/s)**
Air	343
Water	1,483
Steel	5,940
Glass	5,640

23. Calculate Using the table above, determine the total amount of time needed for a sound wave to travel 3.5 km through air and then 100.0 m through water.

24. Predict If the holes of a flute are all covered while playing, then all uncovered, what happens to the length of the vibrating air column? What happens to the pitch of the note?

25. Identify Variables and Controls Describe an experiment to demonstrate that sound is diffracted.

26. Interpret Scientific Illustrations The picture below shows pan pipes. How are different notes produced by blowing on pan pipes?

Performance Activities

27. Recital Perform a short musical piece on an instrument. Explain how your actions changed the notes that were produced.

28. Pamphlet Make a pamphlet describing how a hearing aid works.

29. Interview Interview several people over 65 with some form of hearing loss. Create a table that shows the age of each person and how their hearing has changed with age.

Applying Math

30. Beats Two flutes are playing at the same time. One flute plays a note with a frequency of 524 Hz. If two beats per second are heard, what are the possible frequencies the other flute is playing?

31. Overtones Make a table showing the first three overtones of C, which has a frequency of 262 Hz, and G, which has a frequency of 392 Hz.

Part 1 Multiple Choice

Record your answers on the answer sheet provided by your teacher or on a sheet of paper.

1. In which of the following materials does sound travel the fastest?
   A. empty space   C. air
   B. water   D. steel

2. How can the pitch of the sound made by a guitar string be lowered?
   A. by shortening the part of the string that vibrates
   B. by tightening the string
   C. by replacing the string with a thicker string
   D. by plucking the string harder

Use the figure below to answer questions 3 and 4.

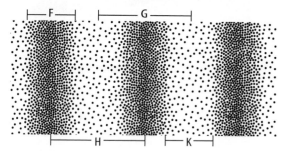

3. What part of the wave is shown at F?
   A. rarefaction   C. wavelength
   B. compression   D. amplitude

4. What part of the wave is shown at H?
   A. rarefaction   C. wavelength
   B. compression   D. amplitude

5. What happens to the particles of matter when a compressional wave moves through the matter?
   A. The particles do not move.
   B. The particles move back and forth along the wave direction.
   C. The particles move back and forth and are carried along with the wave.
   D. The particles move at right angles to the direction the wave travels.

6. If you were on a moving train, what would happen to the pitch of a bell at a crossing as you approached and then passed by the crossing?
   A. It would seem higher, then lower.
   B. It would remain the same.
   C. It would seem lower and then higher.
   D. It would keep getting lower.

Use the figure below to answer questions 7 and 8.

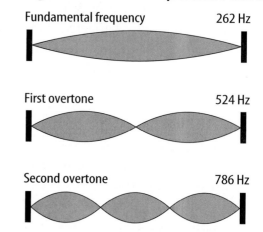

7. How are the overtone frequencies of any vibrating object related to the fundamental frequency of vibration?
   A. They are multiples of the fundamental.
   B. They are not related to the fundamental.
   C. They equal twice the fundamental.
   D. They are lower than the fundamental.

8. Which of the following is the frequency of the third overtone?
   A. 1,572 Hz   C. 1,048 Hz
   B. 1,000 Hz   D. 786 Hz

9. Which of the following is NOT related to the amplitude of a sound wave?
   A. energy carried by the wave
   B. loudness of a sound
   C. pitch of a sound
   D. how spread out the particles are in the compressions and rarefactions

Part 2 Short Response/Grid In

Record your answers on the answer sheet provided by your teacher or on a sheet of paper.

10. What is the difference between diffracted sound waves and echoes?

Use the figure below to answer questions 11–13.

Speed of Sound in Different Materials	
Material	**Speed of sound (m/s)**
Air	343
Water	1,483
Steel	5,940

11. A fish locator sends out a pulse of ultrasound and measures the time needed for the sound to travel to a school of fish and back to the boat. If the fish are 16 m below the boat, how long would it take sound to make the round trip in the water?

12. Suppose you are at a baseball game 150 m from home plate. How long after the batter hits the ball do you hear the sound?

13. A friend drops a stone on a steel railroad track. If the sound made by the stone hitting the track reaches you in 0.8 s, how far away is your friend?

14. Why do different objects produce different sounds when they are struck?

15. Explain how one vibrating tuning fork could make a second tuning fork also vibrate. What is this an example of?

Test-Taking Tip

Notice Units Read carefully and make note of the units used in any measurement.

Question 13 Notice the units used for time in the question and the units for speed given in the table.

Part 3 Open Ended

Record your answers on a sheet of paper.

16. Why do different musical instruments sound different even when they play a note with the same pitch?

17. Compare the way a drum and a flute produce sound waves. What acts as a resonator in each instrument?

18. Would sound waves traveling through the outer ear travel faster or slower than those traveling through the inner ear? Explain.

Use the figure below to answer questions 19.

19. Describe how the process shown in the figure can be used to map the ocean floor.

20. When a sound wave passes through an opening, what does the amount of diffraction depend on?

21. Describe how a cello produces and amplifies sounds.

22. People who work on the ground near jet runways are required to wear ear protection. Explain why this is necessary.

23. Bats use ultrasound when they echolocate prey. If ultrasound waves bounce off an insect that is flying away from the bat, how would the frequency of the wave be affected? What is this effect called?

Electromagnetic Waves

The BIG Idea

Electromagnetic waves are made of changing electric and magnetic fields.

Looking Through You

This color-enhanced X-ray image of a human shoulder and ribcage was made possible by electromagnetic waves. These waves are used to transmit the programs you watch on TV and they make your skin feel warm when you sit in sunlight. In fact, no matter where you go, you are always surrounded by electromagnetic waves.

Science Journal Describe how sitting in sunlight makes you feel. How can sunlight affect your skin?

Start-Up Activities

Detecting Invisible Waves

Light is a type of wave called an electromagnetic wave. You see light every day, but visible light is only one type of electromagnetic wave. Other electromagnetic waves are all around you, but you cannot see them. How can you detect electromagnetic waves that can't be seen with your eyes?

1. Cut a slit 2 cm long and 0.25 cm wide in the center of a sheet of black paper.

2. Cover a window that is in direct sunlight with the paper.

3. Position a glass prism in front of the light coming through the slit so it makes a visible spectrum on the floor or table.

4. Place one thermometer in the spectrum and a second thermometer just beyond the red light.

5. Measure the temperature in each region after 5 min.

6. **Think Critically** Write a paragraph in your Science Journal comparing the temperatures of the two regions and offer an explanation for the observed temperatures.

FOLDABLES
Study Organizer

Electromagnetic Waves Make the following Foldable to help you understand the electromagnetic spectrum.

STEP 1 Collect 4 sheets of paper and layer them about 1 cm apart vertically. Keep the edges level.

STEP 2 Fold up the bottom edges of the paper to form 8 equal tabs.

STEP 3 Fold the papers and crease well to hold the tabs in place. Staple along the fold. Label each tab as indicated below.

Sequence Turn your Foldable so the staples are at the top. Label the tabs, in order from top to bottom, *Electromagnetic Spectrum, Radio Waves, Microwaves, Infrared Rays, Visible Light, Ultraviolet Light, X Rays,* and *Gamma Rays.* As you read, write facts you learn about each topic under the appropriate tab.

Preview this chapter's content and activities at
booko.msscience.com

Get Ready to Read

Visualize

1 **Learn It!** Visualize by forming mental images of the text as you read. Imagine how the text descriptions look, sound, feel, smell, or taste. Look for any pictures or diagrams on the page that may help you add to your understanding.

2 **Practice It!** Read the following paragraph. As you read, use the underlined details to form a picture in your mind.

> A radar station sends out radio waves that bounce off an object, such as an airplane. Electronic equipment measures the time it takes for the radio waves to travel to the plane, be reflected, and return. Because the speed of the radio waves is known, the distance to the airplane can be determined from the measured time.
>
> — *from page 73*

Based on the description above, try to visualiz how radar is used. Now look at the diagram on page 527.
- How closely does it match your mental picture?
- Reread the passage and look at the diagram again. Did your ideas change?
- Compare your image with what others in your class visualized.

3 **Apply It!** Read the chapter and list three subjects you were able to visualize. Make a rough sketch showing what you visualized.

Reading Tip

Forming your own mental images will help you remember what you read.

Target Your Reading

Use this to focus on the main ideas as you read the chapter.

(1) Before you read the chapter, respond to the statements below on your worksheet or on a numbered sheet of paper.

- Write an **A** if you **agree** with the statement.
- Write a **D** if you **disagree** with the statement.

(2) After you read the chapter, look back to this page to see if you've changed your mind about any of the statements.

- If any of your answers changed, explain why.
- Change any false statements into true statements.
- Use your revised statements as a study guide.

Science Online

Print out a worksheet of this page at booko.msscience.com

Before You Read A or D		Statement	After You Read A or D
	1	An electromagnetic wave is a mechanical wave.	
	2	An electromagnetic wave is produced by a moving particle.	
	3	A moving electric charge is surrounded by an electric field and a magnetic field.	
	4	All electromagnetic waves travel at the same speed in empty space.	
	5	Radio waves have the highest frequencies in the electromagnetic spectrum.	
	6	The Sun emits mostly ultraviolet waves.	
	7	Most telecommunication devices use radio waves to transmit information.	
	8	A loudspeaker converts sound waves into electromagnetic waves.	
	9	Radio waves can travel through Earth.	

The Nature of Electromagnetic Waves

Waves in Space

On a clear day you feel the warmth in the Sun's rays, and you see the brightness of its light. Energy is being transferred from the Sun to your skin and eyes. Who would guess that the way in which this energy is transferred has anything to do with radios, televisions, microwave ovens, or the X-ray pictures that are taken by a doctor or dentist? Yet the Sun and the devices shown in **Figure 1** use the same type of wave to move energy from place to place.

Transferring Energy   A wave transfers energy from one place to another without transferring matter. How do waves transfer energy? Waves, such as water waves and sound waves, transfer energy by making particles of matter move. The energy is passed along from particle to particle as they collide with their neighbors. Mechanical waves are the types of waves that use matter to transfer energy.

However, mechanical waves can't travel in the almost empty space between Earth and the Sun. So how can a wave transfer energy from the Sun to Earth? A different type of wave, called an electromagnetic wave, carries energy from the Sun to Earth. An **electromagnetic wave** is a wave that can travel through empty space or through matter and is produced by charged particles that are in motion.

Figure 1   Getting a dental X ray or talking on a cell phone uses energy carried by electromagnetic waves.

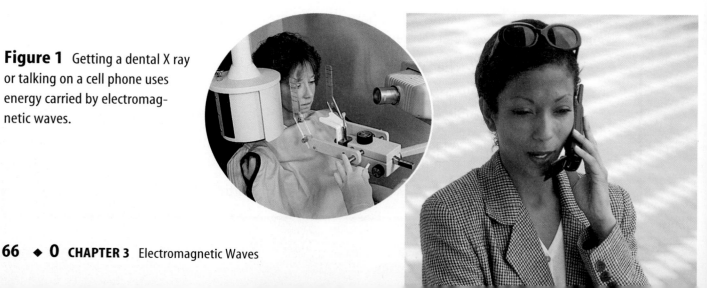

Force
due to
gravity

Earth's gravitational field extends out through space, exerting a force on all masses.
Determine *whether the forces exerted by Earth's gravitational field are attractive or repulsive.*

Forces and Fields

An electromagnetic wave is made of two parts—an electric field and a magnetic field. These fields are force fields. A force field enables an object to exert forces on other objects, even though they are not touching. Earth is surrounded by a force field called the gravitational field. This field exerts the force of gravity on all objects that have mass.

Reading Check *What force field surrounds Earth?*

How does Earth's force field work? If you throw a ball in the air as high as you can, it always falls back to Earth. At every point along the ball's path, the force of gravity pulls down on the ball, as shown in **Figure 2.** In fact, at every point in space above or at Earth's surface, a ball is acted on by a downward force exerted by Earth's gravitational field. The force exerted by this field on a ball could be represented by a downward arrow at any point in space. The diagram above shows this force field that surrounds Earth and extends out into space. It is Earth's gravitational field that causes the Moon to orbit Earth.

Magnetic Fields You know that magnets repel and attract each other even when they aren't touching. Two magnets exert a force on each other when they are some distance apart because each magnet is surrounded by a force field called a magnetic field. Just as a gravitational field exerts a force on a mass, a magnetic field exerts a force on another magnet and on magnetic materials. Magnetic fields cause other magnets to line up along the direction of the magnetic field.

Topic: Force Fields
Visit to booko.msscience.com for Web links to information about Earth's gravitational and magnetic force fields.

Activity Write a paragraph comparing and contrasting the two force fields.

Figure 3 Force fields surround all magnets and electric charges.

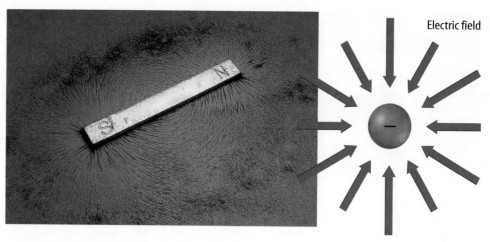

A magnetic field surrounds all magnets. The magnetic field exerts a force on iron filings, causing them to line up with the field.

Electric field

The electric field around an electric charge extends out through space, exerting forces on other charged particles.

Electric Fields Recall that atoms contain protons, neutrons, and electrons. Protons and electrons have a property called electric charge. The two types of electric charge are positive and negative. Protons have positive charge and electrons have negative charge.

Just as a magnet is surrounded by a magnetic field, a particle that has electric charge, such as a proton or an electron, is surrounded by an electric field, as shown in **Figure 3.** The electric field is a force field that exerts a force on all other charged particles that are in the field.

Making Electromagnetic Waves

An electromagnetic wave is made of electric and magnetic fields. How is such a wave produced? Think about a wave on a rope. You can make a wave on a rope by shaking one end of the rope up and down. Electromagnetic waves are produced by charged particles, such as electrons, that move back and forth or vibrate.

A charged particle always is surrounded by an electric field. But a charged particle that is moving also is surrounded by a magnetic field. For example, electrons are flowing in a wire that carries an electric current. As a result, the wire is surrounded by a magnetic field, as shown in **Figure 4.** So a moving charged particle is surrounded by an electric field and a magnetic field.

Figure 4 Electrons moving in a wire produce a magnetic field in the surrounding space. This field causes iron filings to line up with the field.

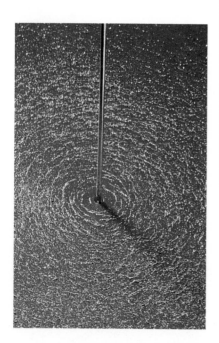

Magnetic field

Moving electrons

Producing Waves When you shake a rope up and down, you produce a wave that moves away from your hand. As a charged particle vibrates by moving up and down or back and forth, it produces changing electric and magnetic fields that move away from the vibrating charge in many directions. These changing fields traveling in many directions form an electromagnetic wave. **Figure 5** shows how the electric and magnetic fields change as they move along one direction.

Properties of Electromagnetic Waves

Like all waves, an electromagnetic wave has a frequency and a wavelength. You can create a wave on a rope when you move your hand up and down while holding the rope. Look at **Figure 5.** Frequency is how many times you move the rope through one complete up and down cycle in 1 s. Wavelength is the distance from one crest to the next or from one trough to the next.

Wavelength and Frequency An electromagnetic wave is produced by a vibrating charged particle. When the charge makes one complete vibration, one wavelength is created, as shown in **Figure 5.** Like a wave on a rope, the frequency of an electromagnetic wave is the number of wavelengths that pass by a point in 1 s. This is the same as the number of times in 1 s that the charged particle makes one complete vibration.

Mini LAB

Observing Electric Fields

Procedure
1. Rub a **hard, plastic comb** vigorously with a **wool sweater or wool flannel shirt.**
2. Turn on a **water faucet** to create the smallest possible continuous stream of water.
3. Hold the comb near the stream of water and observe.

Analysis
1. What happened to the stream of water when you held the comb near it?
2. Explain why the stream of water behaved this way.

Try at Home

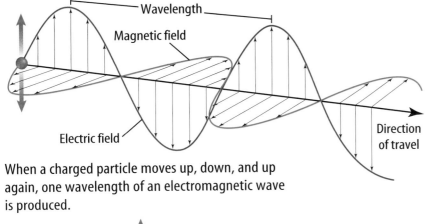

When a charged particle moves up, down, and up again, one wavelength of an electromagnetic wave is produced.

Figure 5 The vibrating motion of an electric charge produces an electromagnetic wave. One complete cycle of vibration produces one wavelength of a wave. **Determine** *the magnetic field when the electric field is zero.*

By shaking the end of a rope down, up, and down again, you make one wavelength.

Alpha Centauri

Figure 6 The light that reaches Earth today from Alpha Centauri left the star more than four years ago.

Radiant Energy The energy carried by an electromagnetic wave is called **radiant energy.** What happens if an electromagnetic wave strikes a charged particle? The electric field part of the wave exerts a force on this particle and causes it to move. Some of the radiant energy carried by the wave is transferred into the energy of motion of the particle.

✔ **Reading Check** *What is radiant energy?*

The amount of energy that an electromagnetic wave carries is determined by the wave's frequency. The higher the frequency of the electromagnetic wave, the more energy it has.

The Speed of Light All electromagnetic waves travel through space at the same speed—about 300,000 km/s. This speed sometimes is called the speed of light. Even though light travels incredibly fast, stars other than the Sun are so far away that it takes years for the light they emit to reach Earth. **Figure 6** shows Alpha Centauri, one of the closest stars to our solar system. This star is more than 40 trillion km from Earth.

---

section ① review

Summary

Force Fields

- A charged particle is surrounded by an electric field that exerts forces on other charged particles.
- A magnet is surrounded by a magnetic field that exerts a force on other magnets.
- A moving charged particle is surrounded by electric and magnetic fields.

Electromagnetic Waves

- The changing electric and magnetic fields made by a vibrating electric charge form an electromagnetic wave.
- Electromagnetic waves carry radiant energy.
- All electromagnetic waves travel at the speed of light, which is about 300,000 km/s in empty space.

Self Check

1. **Describe** how electromagnetic waves are produced.
2. **Compare** the energy carried by high-frequency and low-frequency electromagnetic waves.
3. **Identify** what determines the frequency of an electromagnetic wave.
4. **Compare and contrast** electromagnetic waves with mechanical waves.
5. **Think Critically** Unlike sound waves, electromagnetic waves can travel in empty space. What evidence supports this statement?

Applying Math

6. **Use Ratios** To go from Earth to Mars, light waves take four min and a spacecraft takes four months. To go to the nearest star, light takes four years. How long would it take the spacecraft to go to the nearest star?

Science Online booko.msscience.com/self_check_quiz

The Electromagnetic Spectrum

Electromagnetic Waves

The room you are sitting in is bathed in a sea of electromagnetic waves. These electromagnetic waves have a wide range of wavelengths and frequencies. For example, TV and radio stations broadcast electromagnetic waves that pass through walls and windows. These waves have wavelengths from about 1 m to over 500 m. Light waves that you see are electromagnetic waves that have wavelengths more than a million times shorter than the waves broadcast by radio stations.

Classifying Electromagnetic Waves The wide range of electromagnetic waves with different frequencies and wavelengths forms the **electromagnetic spectrum.** The electromagnetic spectrum is divided into different parts. **Figure 7** shows the electromagnetic spectrum and the names given to the electromagnetic waves in different parts of the spectrum. Even though electromagnetic waves have different names, they all travel at the same speed in empty space—the speed of light. Remember that for waves that travel at the same speed, the frequency increases as the wavelength decreases. So as the frequency of electromagnetic waves increases, their wavelength decreases.

Figure 7 The electromagnetic spectrum consists of electromagnetic waves arranged in order of increasing frequency and decreasing wavelength.

as you read

What You'll Learn
- **Explain** differences among kinds of electromagnetic waves.
- **Identify** uses for different kinds of electromagnetic waves.

Why It's Important
Electromagnetic waves are used to cook food, to send and receive information, and to diagnose medical problems.

Review Vocabulary
spectrum: a continuous series of waves arranged in order of increasing or decreasing wavelength or frequency

New Vocabulary
- electromagnetic spectrum
- radio wave
- infrared wave
- visible light
- ultraviolet radiation
- X ray
- gamma ray

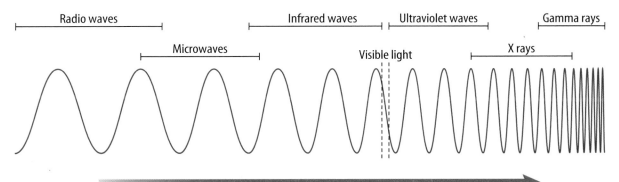

Increasing frequency, decreasing wavelength

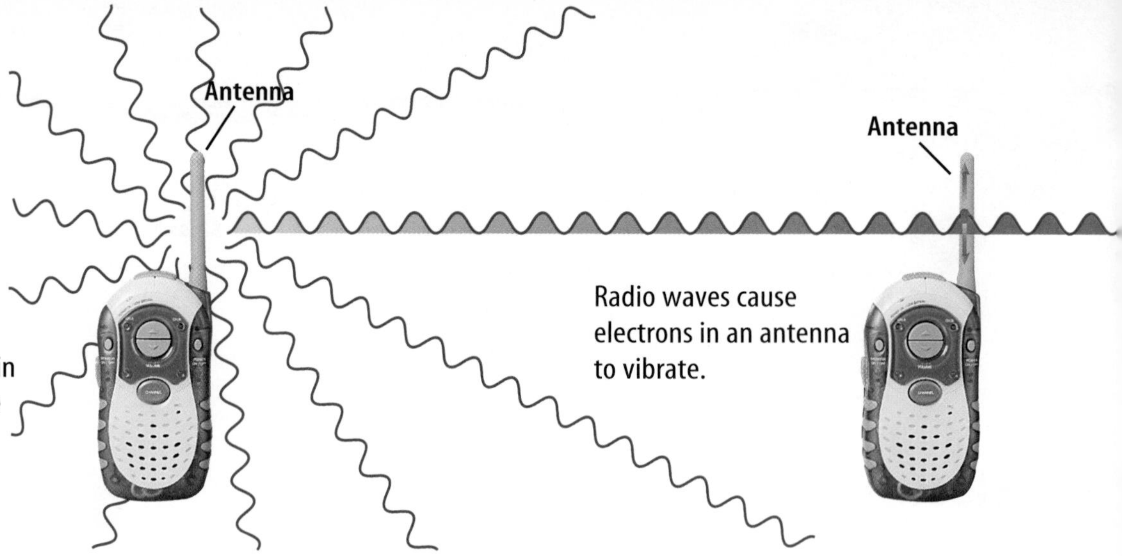

Figure 8 Antennas are used to generate and detect radio waves. **Describe** *some objects that have antennas.*

Antenna

Antenna

Vibrating electrons in an antenna produce radio waves.

Radio waves cause electrons in an antenna to vibrate.

Radio Waves

Electromagnetic waves with wavelengths longer than about 0.001 m are called radio waves. **Radio waves** have the lowest frequencies of all the electromagnetic waves and carry the least energy. Television signals, as well as AM and FM radio signals, are types of radio waves. Like all electromagnetic waves, radio waves are produced by moving charged particles. One way to make radio waves is to make electrons vibrate in a piece of metal, as shown in **Figure 8.** This piece of metal is called an antenna. By changing the rate at which the electrons vibrate, radio waves of different frequencies can be produced that travel outward from the antenna.

Detecting Radio Waves These radio waves can cause electrons in another piece of metal, such as another antenna, to vibrate, as shown in **Figure 8.** As the electrons in the receiving antenna vibrate, they form an alternating current. This alternating current can be used to produce a picture on a TV screen and sound from a loudspeaker. Varying the frequency of the radio waves broadcast by the transmitting antenna changes the alternating current in the receiving antenna. This produces the different pictures you see and sounds you hear on your TV.

Microwaves Radio waves with wavelengths between about 0.3 m and 0.001 m are called microwaves. They have a higher frequency and a shorter wavelength than the waves that are used in your home radio. Microwaves are used to transmit some phone calls, especially from cellular and portable phones. **Figure 9** shows a microwave tower.

Microwave ovens use microwaves to heat food. Microwaves produced inside a microwave oven cause water molecules in your food to vibrate faster, which makes the food warmer.

Figure 9 Towers such as the one shown here are used to send and receive microwaves.

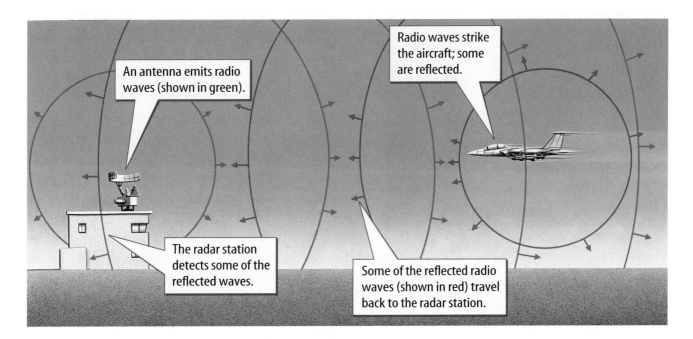

Radar waves strike the aircraft; some are reflected.

An antenna emits radio waves (shown in green).

The radar station detects some of the reflected waves.

Some of the reflected radio waves (shown in red) travel back to the radar station.

Radar You might be familiar with echolocation, in which sound waves are reflected off an object to determine its size and location. Some bats and dolphins use echolocation to navigate and hunt. Radar, an acronym for RAdio Detecting And Ranging, uses electromagnetic waves to detect objects in the same way. Radar was first used during World War II to detect and warn of incoming enemy aircraft.

Reading Check *What does radar do?*

A radar station sends out radio waves that bounce off an object such as an airplane. Electronic equipment measures the time it takes for the radio waves to travel to the plane, be reflected, and return. Because the speed of the radio waves is known, the distance to the airplane can be determined from the measured time.

An example of radar being used is shown in **Figure 10.** Because electromagnetic waves travel so quickly, the entire process takes only a fraction of a second.

Infrared Waves

You might know from experience that when you stand near the glowing coals of a barbecue or the red embers of a campfire, your skin senses the heat and becomes warm. Your skin may also feel warm near a hot object that is not glowing. The heat you are sensing with your skin is from electromagnetic waves. These electromagnetic waves are called **infrared waves** and have wavelengths between about one thousandth and 0.7 millionths of a meter.

Figure 10 Radar stations use radio waves to determine direction, distance, and speed of aircraft.

Observing the Focusing of Infrared Rays

Procedure 🖐️ ✋
1. Place a **concave mirror** 2 m to 3 m away from an **electric heater.** Turn on the heater.
2. Place the palm of your hand in front of the mirror and move it back until you feel heat on your palm. Note the location of the warm area.
3. Move the heater to a new location. How does the warm area move?

Analysis
1. Did you observe the warm area? Where?
2. Compare the location of the warm area to the location of the mirror.

Figure 11 A pit viper hunting in the dark can detect the infrared waves emitted from the warm body of its prey.

Detecting Infrared Waves Electromagnetic waves are emitted by every object. In any material, the atoms and molecules are in constant motion. Electrons in the atoms and molecules also are vibrating, and so they emit electromagnetic waves. Most of the electromagnetic waves given off by an object at room temperature are infrared waves and have a wavelength of about 0.000 01 m, or one hundred thousandth of a meter.

Infrared detectors can detect objects that are warmer or cooler than their surroundings. For example, areas covered with vegetation, such as forests, tend to be cooler than their surroundings. Using infrared detectors on satellites, the areas covered by forests and other vegetation, as well as water, rock, and soil, can be mapped. Some types of night vision devices use infrared detectors that enable objects to be seen in nearly total darkness.

Animals and Infrared Waves Some animals also can detect infrared waves. Snakes called pit vipers, such as the one shown in **Figure 11,** have a pit located between the nostril and the eye that detects infrared waves. Rattlesnakes, copperheads, and water moccasins are pit vipers. These pits help pit vipers hunt at night by detecting the infrared waves their prey emits.

Visible Light

As the temperature of an object increases, the atoms and molecules in the object move faster. The electrons also vibrate faster, and produce electromagnetic waves of higher frequency and shorter wavelength. If the temperature is high enough, the object might glow, as in **Figure 12.** Some of the electromagnetic waves that the hot object is emitting are now detectable with your eyes. Electromagnetic waves you can detect with your eyes are called **visible light.** Visible light has wavelengths between about 0.7 and 0.4 millionths of a meter. What you see as different colors are electromagnetic waves of different wavelengths. Red light has the longest wavelength (lowest frequency), and blue light has the shortest wavelength (highest frequency).

Most objects that you see do not give off visible light. They simply reflect the visible light that is emitted by a source of light, such as the Sun or a lightbulb.

Figure 12 When objects are heated, their electrons vibrate faster. When the temperature is high enough, the vibrating electrons will emit visible light. **Describe** *an object that emits visible light when heated.*

Electromagnetic Waves from the Sun

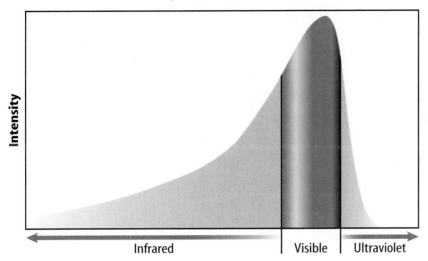

Intensity

Infrared | Visible | Ultraviolet

Figure 13 The Sun emits mainly infrared waves and visible light. Only about 8 percent of the electromagnetic waves emitted by the Sun are ultraviolet radiation. **Identify** *the electromagnetic waves emitted by the Sun that have the highest intensity.*

Ultraviolet Radiation

Ultraviolet radiation is higher in frequency than visible light and has even shorter wavelengths—between 0.4 millionths of a meter and about ten billionths of a meter. Ultraviolet radiation has higher frequencies than visible light and carries more energy. The radiant energy carried by an ultraviolet wave can be enough to damage the large, fragile molecules that make up living cells. Too much ultraviolet radiation can damage or kill healthy cells.

Figure 13 shows the intensity of electromagnetic waves emitted by the Sun. Too much exposure to the Sun's ultraviolet waves can cause sunburn. Exposure to these waves over a long period of time can lead to early aging of the skin and possibly skin cancer. You can reduce the amount of ultraviolet radiation you receive by wearing sunglasses and sunscreen, and staying out of the Sun when it is most intense.

Reading Check *Why can too much exposure to the Sun be harmful?*

Figure 14 Sterilizing devices, such as this goggle sterilizer, use ultraviolet waves to kill organisms on the equipment.

Beneficial Uses of UV Radiation A few minutes of exposure each day to ultraviolet radiation from the Sun enables your body to produce the vitamin D it needs. Most people receive that amount during normal activity. The body's natural defense against too much ultraviolet radiation is to tan. However, a tan can be a sign that overexposure to ultraviolet radiation has occurred.

Because ultraviolet radiation can kill cells, it is used to disinfect surgical equipment in hospitals. In some chemistry labs, ultraviolet rays are used to sterilize goggles, as shown in **Figure 14.**

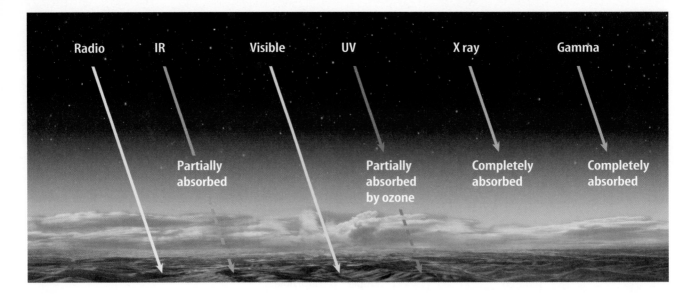

Radio   IR   Visible   UV   X ray   Gamma

Partially absorbed   Partially absorbed by ozone   Completely absorbed   Completely absorbed

Figure 15 Earth's atmosphere serves as a shield to block some types of electromagnetic waves from reaching Earth's surface.

INTEGRATE Life Science

Body Temperature Warm-blooded animals, such as mammals, produce their own body heat. Cold-blooded animals, such as reptiles, absorb heat from the environment. Brainstorm the possible advantages of being either warm-blooded or cold-blooded. Which animals would be easier for a pit viper to detect?

The Ozone Layer Much of the ultraviolet radiation arriving at Earth is absorbed in the upper atmosphere by ozone, as shown in **Figure 15.** Ozone is a molecule that has three oxygen atoms and is formed high in Earth's atmosphere.

Chemical compounds called CFCs, which are used in air conditioners and refrigerators, can react with ozone molecules and break them apart. There is evidence that these reactions play a role in forming the seasonal reduction in ozone over Antarctica, known as the ozone hole. To prevent this, the use of CFC's is being phased out.

Ultraviolet radiation is not the only type of electromagnetic wave absorbed by Earth's atmosphere. Higher energy waves of X rays and gamma rays also are absorbed. The atmosphere is transparent to radio waves and visible light and partially transparent to infrared waves.

X Rays and Gamma Rays

Ultraviolet rays can penetrate the top layer of your skin. **X rays,** with an even higher frequency than ultraviolet rays, have enough energy to go right through skin and muscle. A shield made from a dense metal, such as lead, is required to stop X rays.

Gamma rays have the highest frequency and, therefore, carry the most energy. Gamma rays are the hardest to stop. They are produced by changes in the nuclei of atoms. When protons and neutrons bond together in nuclear fusion or break apart from each other in nuclear fission, enormous quantities of energy are released. Some of this energy is released as gamma rays.

Just as too much ultraviolet radiation can hurt or kill cells, too much X-ray or gamma radiation can have the same effect. Because the energy of X rays and gamma rays is greater, the exposure that is needed to cause damage is much less.

Using High-Energy Electromagnetic Radiation The fact that X rays can pass through the human body makes them useful for medical diagnosis, as shown in **Figure 16.** X rays pass through the less dense tissues in skin and other organs. These X rays strike a film, creating a shadow image of the denser tissues. X-ray images help doctors detect injuries and diseases, such as broken bones and cancer. A CT scanner uses X rays to produce images of the human body as if it had been sliced like a loaf of bread.

Although the radiation received from getting one medical or dental X ray is not harmful, the cumulative effect of numerous X rays can be dangerous. The operator of the X-ray machine usually stands behind a shield to avoid being exposed to X rays. Lead shields or aprons are used to protect the parts of the patient's body that are not receiving the X rays.

Using Gamma Rays Although gamma rays are dangerous, they also have beneficial uses, just as X rays do. A beam of gamma rays focused on a cancerous tumor can kill the tumor. Gamma radiation also can kill disease-causing bacteria in food. More than 1,000 Americans die each year from *Salmonella* bacteria in poultry and *E. coli* bacteria in meat. Although gamma radiation has been used since 1963 to kill bacteria in food, this method is not widely used in the food industry.

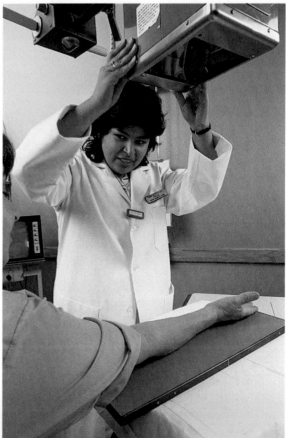

Figure 16 Dense tissues such as bone absorb more X rays than softer tissues do. Consequently, dense tissues leave a shadow on an X ray film that can be used to diagnose medical and dental conditions.

Astronomy with Different Wavelengths

Some astronomical objects produce no visible light and can be detected only through the infrared and radio waves they emit. Some galaxies emit X rays from regions that do not emit visible light. Studying stars and galaxies like these using only visible light would be like looking at only one color in a picture. **Figure 17** shows how different electromagnetic waves can be used to study the Sun.

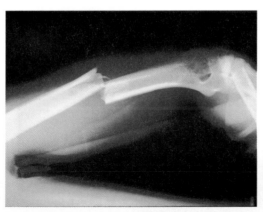

Figure 17

For centuries, astronomers studied the universe using only the visible light coming from planets, moons, and stars. But many objects in space also emit X rays, ultraviolet and infrared radiation, and radio waves. Scientists now use telescopes that can detect these different types of electromagnetic waves. As these images of the Sun reveal, the new tools are providing more information of objects in the universe.

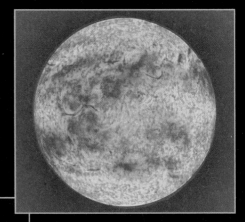

▲ **INFRARED RADIATION** An infrared telescope reveals that the Sun's surface temperature is not uniform. Some areas are hotter than others.

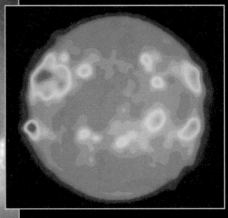

▲ **RADIO WAVES** Radio telescopes detect radio waves given off by the Sun, which have much longer wavelengths than visible light.

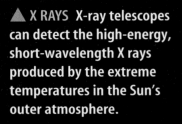

▲ **X RAYS** X-ray telescopes can detect the high-energy, short-wavelength X rays produced by the extreme temperatures in the Sun's outer atmosphere.

▶ **ULTRAVIOLET RADIATION** Telescopes sensitive to ultraviolet radiation—electromagnetic waves with shorter wavelengths than visible light—can "see" the Sun's outer atmosphere.

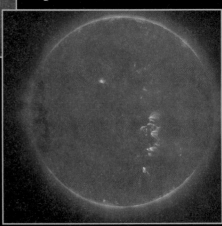

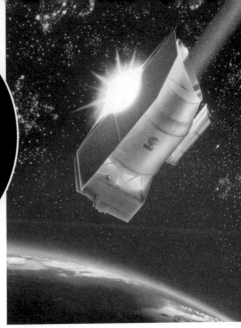

 **Satellite Observations** Recall from **Figure 15** that Earth's atmosphere blocks X rays, gamma rays, most ultraviolet rays, and some infrared rays. However, telescopes in orbit above Earth's atmosphere can detect the electromagnetic waves that can't pass through the atmosphere. **Figure 18** shows three such satellites—the Extreme Ultraviolet Explorer (EUVE), the Chandra X-Ray Observatory, and the Infrared Space Observatory (ISO).

Figure 18 Launching satellite observatories above Earth's atmosphere is the only way to see the universe at electromagnetic wavelengths that are absorbed by Earth's atmosphere.

section 2 review

Summary

Radio Waves

- Radio waves have wavelengths longer than about 0.3 m.

Infrared Waves and Visible Light

- Infrared waves have wavelengths between about one thousandth and 0.7 millionths of a meter.

- The wavelengths of infrared waves emitted by an object get shorter as the object's temperature increases.

- Visible light waves have wavelengths between about 0.7 and 0.4 millionths of a meter.

Ultraviolet Waves, X Rays, and Gamma Rays

- Ultraviolet radiation has wavelengths between about 0.4 millionths of a meter and 10 billionths of a meter.

- Prolonged exposure to ultraviolet waves from the Sun can cause skin damage.

- X rays and gamma rays are the most energetic electromagnetic waves.

Self Check

1. **Explain** why ultraviolet radiation is more damaging to living cells than infrared waves.

2. **Compare and contrast** X rays and gamma rays.

3. **Describe** how infrared detectors on satellites can be used to obtain information about the location of vegetation on Earth's surface.

4. **Explain** why X rays and gamma rays coming from space do not reach Earth's surface.

5. **Explain** how the energy of electromagnetic waves change as the wavelength of the waves increase.

6. **Think Critically** Why does the Sun emit mostly infrared waves and visible light, and Earth emits infrared waves?

Applying Skills

7. **Make a table** listing five objects in your home that produce electromagnetic waves. In another column, list next to each object the type of electromagnetic wave or waves produced. In a third column describe each object's use.

Prisms  f Light

Do you know what light is? Many would answer that light is what you turn on to see at night. However, white light is made of many different frequencies of the electromagnetic spectrum. A prism can separate white light into its different frequencies. You see different frequencies of light as different colors. What colors do you see when light passes through a prism?

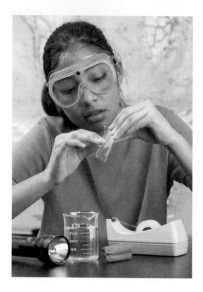

◉ Real-World Question

What happens to visible light as it passes through a prism?

Goals
- **Construct** a prism and observe the different colors that are produced.
- **Infer** how the bending of light waves depends on their wavelength.

Materials
microscope slides (3)   flashlight
transparent tape   water
clay   *prism

*Alternate materials

Safety Precautions

◉ Procedure

1. Carefully tape the three slides together on their long sides so they form a long prism.

2. Place one end of the prism into a softened piece of clay so the prism is standing upright.

3. Fill the prism with water and put it on a table that is against a dark wall.

4. Shine a flashlight beam through the prism so the light becomes visible on the wall.

◉ Conclude and Apply

1. **List** the order of the colors you saw on the wall.

2. **Describe** how the position of the colors on the wall changes as you change the direction of the flashlight beam.

3. **Describe** how the order of colors on the wall changes as you change the direction of the flashlight beam.

4. **Infer** which color light waves have changed direction, or have been bent, the most after passing through the prism. Which color has been bent the least?

5. **Infer** how the bending of a light wave depends on its wavelength.

𝒞ommunicating
Your Data

Compare your conclusions with those of other students in your class. **For more help, refer to the** Science Skill Handbook.

Using Electromagnetic Waves

Telecommunications

In the past week, have you spoken on the phone, watched television, done research on the Internet, or listened to the radio? Today you can talk to someone far away or transmit and receive information over long distances almost instantly. Thanks to telecommunications, the world is becoming increasingly connected through the use of electromagnetic waves.

Using Radio Waves

Radio waves usually are used to send and receive information over long distances. Using radio waves to communicate has several advantages. For example, radio waves pass through walls and windows easily. Radio waves do not interact with humans, so they are not harmful to people like ultraviolet rays or X rays are. So most telecommunication devices, such as TVs, radios, and telephones, use radio waves to transmit information such as images and sounds. **Figure 19** shows how radio waves can be used to transmit information—in this case transmitting information that enables sounds to be reproduced at a location far away.

as you read

What **You'll Learn**

■ **Describe** different ways of using electromagnetic waves to communicate.
■ **Compare and contrast** AM and FM radio signals.

Why **It's Important**

Using elecromagnetic waves to communicate enables you to contact others worldwide.

🔎 **Review Vocabulary**
satellite: a natural or artificial object that orbits a planet

New Vocabulary
● carrier wave
● Global Positioning System

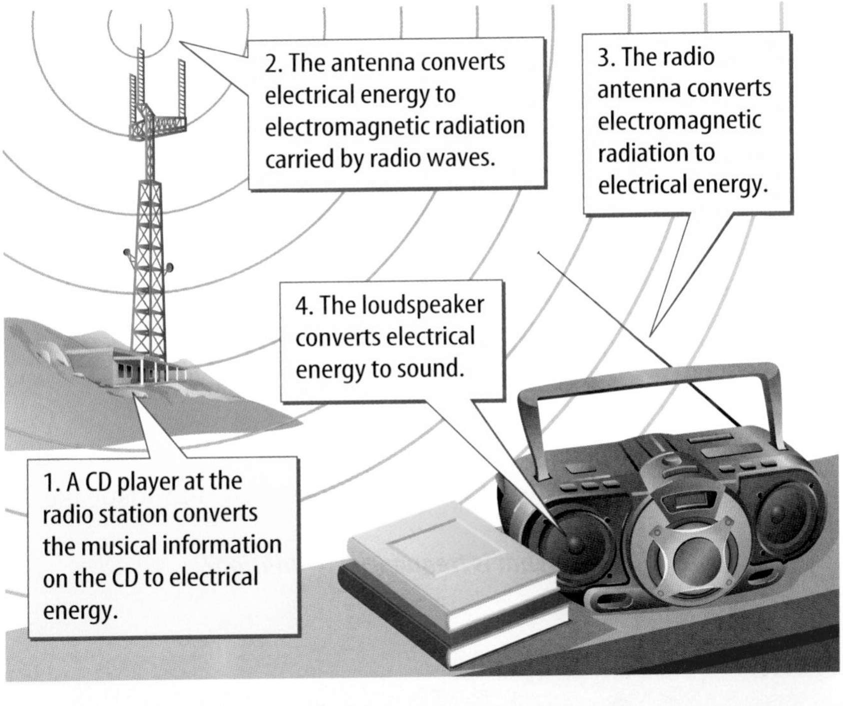

2. The antenna converts electrical energy to electromagnetic radiation carried by radio waves.

3. The radio antenna converts electromagnetic radiation to electrical energy.

4. The loudspeaker converts electrical energy to sound.

1. A CD player at the radio station converts the musical information on the CD to electrical energy.

Figure 19 Radio waves are used to transmit information that can be converted to other forms of energy, such as electrical energy and sound.

Figure 20 A signal can be carried by a carrier wave in two ways—amplitude modulation or frequency modulation.

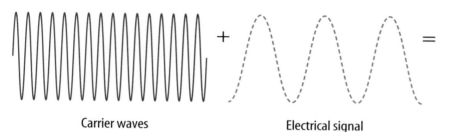

Carrier waves

+

Electrical signal

=

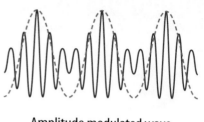

Amplitude modulated wave

or

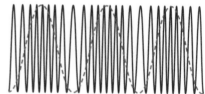

Frequency modulated wave

Pulsars and Little Green Men A type of collapsed star called a pulsar emits pulses of radio waves at extremely regular intervals. Pulsars were first discovered by Jocelyn Bell-Burnell and Anthony Hewish in 1967. Puzzled by a regular sequence of radio pulses they detected, they considered the possibility that the pulses might be coming from an alien civilization. They jokingly labeled the pulses LGMs, for "little green men." Soon other signals were detected that proved the pulses were coming from collapsed stars. Research the role Jocelyn Bell-Burnell played in the discovery of pulsars.

Radio Transmission How is information, such as images or sounds, broadcast by radio waves? Each radio and television station is assigned a particular frequency at which it broadcasts radio waves. The radio waves broadcast by a station at its assigned frequency are the **carrier waves** for that station. To listen to a station you tune your radio or television to the frequency of the station's carrier waves. To carry information on the carrier wave, either the amplitude or the frequency of the carrier wave is changed, or modulated.

Amplitude Modulation The letters *AM* in AM radio stand for amplitude modulation, which means that the amplitude of the carrier wave is changed to transmit information. The original sound is transformed into an electrical signal that is used to vary the amplitude of the carrier wave, as shown in **Figure 20.** Note that the frequency of the carrier wave doesn't change—only the amplitude changes. An AM receiver tunes to the frequency of the carrier wave. In the receiver, the varying amplitude of the carrier waves produces an electrical signal. The radio's loudspeaker uses this electric signal to produce the original sound.

Frequency Modulation FM radio works in much the same way as AM radio, but the frequency instead of the amplitude is modulated, as shown in **Figure 20.** An FM receiver contains electronic components that use the varying frequency of the carrier wave to produce an electric signal. As in an AM radio, this electric signal is converted into sound waves by a loudspeaker.

 **Reading Check** *What is frequency modulation?*

Telephones

A telephone contains a microphone in the mouthpiece that converts a sound wave into an electric signal. The electric signal is carried through a wire to the telephone switching systems. There, the signal might be sent through other wires or be converted into a radio or microwave signal for transmission through the air. The electric signal also can be converted into a light wave for transmission through fiber-optic cables.

At the receiving end, the signal is converted back to an electric signal. A speaker in the earpiece of the phone changes the electric signal into a sound wave.

 **Reading Check** *What device converts sound into an electric signal?*

Applying Math — Solve a Simple Equation

WAVELENGTH OF AN FM STATION You are listening to an FM radio station with a frequency of 94.9 MHz, which equals 94,900,000 Hz. What is the wavelength of these radio waves. Use the wave speed equation $v = \lambda f$, and assume the waves travel at the speed of light, 300,000.0 km/s.

Solution

1 *This is what you know:*
- frequency: f = 94,900,000 Hz
- wave speed: v = 300,000.0 km/s

2 *This is what you need to find:*
- wavelength: λ = ? m

3 *This is the procedure you need to use:*

Solve the wave equation for wavelength, λ, by dividing each side by the frequency, f. Then substitute the known values for frequency and wave speed into the equation you derived:

$$\lambda = \frac{v}{f} = \frac{300{,}000.0 \text{ km/s}}{94{,}900{,}000 \text{ Hz}} = \frac{300{,}000.0 \text{ km } 1/s}{94{,}900{,}000 \; 1/s}$$

$$= 0.00316 \text{ km} = 0.00316 \text{ km} \times (1{,}000 \text{ m/km})$$

$$= 3.16 \text{ m}$$

4 *Check your answer:*

Multiply your answer by the given frequency. The result should be the given wave speed.

Practice Problems

1. Your friend is listening to an AM station with a frequency of 1,520 kHz. What is the wavelength of these radio waves?

2. What is the frequency of the radio waves broadcast by an AM station if the wave length of the radio waves is 500.0 m?

Science Online

For more practice, visit booko.msscience.com/ math_practice

Figure 21 Cordless and cell phones use radio waves to communicate between a mobile phone and a base station.

B Cell phones communicate with a base station that can be several kilometers away, or more.

A A cordless phone can be used more than 0.5 km from its base station.

Remote Phones A telephone does not have to transmit its signal through wires. In a cordless phone, the electrical signal produced by the microphone is transmitted through an antenna in the phone to the base station. **Figure 21A** shows how incoming signals are transmitted from the base station to the phone. A cellular phone communicates with a base station that can be many kilometers away. The base station uses a large antenna, as shown in **Figure 21B,** to communicate with the cell phone and with other base stations in the cell phone network.

Pagers The base station also is used in a pager system. When you dial a pager, the signal is sent to a base station. From there, an electromagnetic signal is sent to the pager. The pager beeps or vibrates to indicate that someone has called. With a touch-tone phone, you can transmit numeric information, such as your phone number, which the pager will receive and display.

Communications Satellites

How do you send information to the other side of the world? Radio waves can't be sent directly through Earth. Instead, radio signals are sent to satellites. The satellites can communicate with other satellites or with ground stations. Some communications satellites are in geosynchronous orbit, meaning each satellite remains above the same point on the ground.

Science Online

Topic: Satellite Communications
Visit booko.msscience.com for Web links to information about how satellites are used in around-the-world communications.

Activity Create a table listing satellites from several countries, their names and their communications function.

The Global Positioning System

Satellites also are used as part of the **Global Positioning System,** or GPS. GPS is used to locate objects on Earth. The system consists of satellites, ground-based stations, and portable units with receivers, as illustrated in **Figure 22.**

A GPS receiver measures the time it takes for radio waves to travel from several satellites to the receiver. This determines the distance to each satellite. The receiver then uses this information to calculate its latitude, longitude, and elevation. The accuracy of GPS receivers ranges from a few hundred meters for handheld units, to several centimeters for units that are used to measure the movements of Earth's crust.

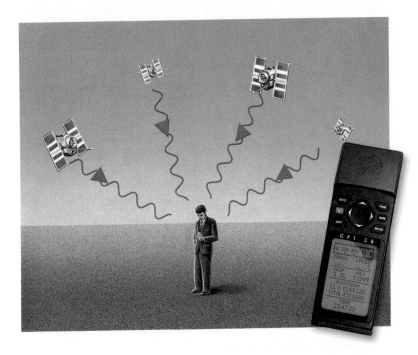

Figure 22 The signals broadcast by GPS satellites enable portable, handheld receivers to determine the position of an object or person.

section 3 review

Summary

Using Radio Waves

- Radio waves are used for communication because they can pass through most objects.
- Amplitude modulation transmits information by modifying the amplitude of a carrier wave.
- Frequency modulation transmits information by modifying the frequency of a carrier wave.

Cordless Phones and Cell Phones

- Cordless phones use radio waves to transmit signals between the base and the handset.
- Cellular phones use radio waves to transmit signals between the phone and cell phone radio towers.

Communications Satellites

- Communications satellites in geosynchronous orbits relay radio signals from one part of the world to another.
- The Global Positioning System uses radio waves to enable a user to accurately determine their position on Earth's surface.

Self Check

1. **Describe** how a cordless phone is different from a cell phone.
2. **Explain** how a communications satellite is used.
3. **Describe** the types of information a GPS receiver provides.
4. **Describe** how an AM radio signal is used to transmit information.
5. **Think Critically** Explain why ultraviolet waves are not used to transmit signals to and from communications satellites.

Applying Skills

6. **Make an events chain** showing the sequence of energy transformations that occur when live music is broadcast by a radio station and played by a radio.
7. **Make a Diagram** showing how geosynchronous satellites and ground stations could be used to send information from you to someone on the other side of Earth.

Spectrum Inspection

◉ *Real-World Question*

You've heard the term "red-hot" used to describe something that is unusually hot. When a piece of metal is heated it may give off a red glow or even a yellow glow. All objects emit electromagnetic waves. How do the wavelengths of these waves depend on the temperature of the object?

◉ *Form a Hypothesis*

The brightness of a lightbulb increases as its temperature increases. Form a hypothesis describing how the wavelengths emitted by a lightbulb will change as the brightness of a lightbulb changes.

◉ *Test Your Hypothesis*

Make a Plan

1. **Decide** how you will determine the effect of lightbulb brightness on the colors of light that are emitted.

2. As shown in the photo at the right, you will look toward the light through the diffraction grating to detect the colors of light emitted by the bulb. The color spectrum will appear to the right and to the left of the bulb.

3. **List** the specific steps you will need to take to test your hypothesis. Describe precisely what you will do in each step. Will you first test the bulb at a bright or dim setting? How many settings will you test? (Try at least three.) How will you record your observations in an organized way?

4. **List** the materials you will need for your experiment. Describe exactly how and in which order you will use these materials.

5. **Identify** any constants and variables in your experiment.

Follow Your Plan

1. Make sure your teacher approves your plan before you start.

2. **Perform** your experiment as planned.

3. While doing your experiment, write down any observations you make in your Science Journal.

▶ *Analyze Your Data*

1. Use the colored pencils to draw the color spectrum emitted by the bulb at each brightness.

2. Which colors appeared as the bulb became brighter? Did any colors disappear?

3. How did the wavelengths emitted by the bulb change as the bulb became brighter?

4. **Infer** how the frequencies emitted by the lightbulb changed as it became hotter.

▶ *Conclude and Apply*

1. **Infer** If an object becomes hotter, what happens to the wavelengths it emits?

2. How do the wavelengths that the bulb emits change if it is turned off?

3. **Infer** from your results whether red stars or yellow stars are hotter.

*C*ommunicating
Your Data

Compare your results with others in your class. How many different colors were seen?

Hedy Lamarr, actor and inventor

Hopping the Frequencies

Ringgggg. There it is—that familiar beep! Out come all the cellular phones. At any given moment, a million wireless signals are flying through the air—and not just cell phone signals. With radio and television signals, Internet data, and even Global Positioning System information, the air seems like a pretty crowded place. How does a cellular phone pick out its own signal from among the clutter? The answer lies in a concept developed in 1940 by Hedy Lamarr.

Lamarr was born in Vienna, Austria. In 1937, she left Austria to escape Hitler's invading Nazi army. She also left to pursue a career as an actor. And she became a famous movie star.

In 1940, Lamarr came up with an idea to keep radio signals that guided torpedoes from being jammed. Her idea, called frequency hopping, involved breaking the radio signal that was guiding the torpedo into tiny parts and rapidly changing their frequency. The enemy would not be able to keep up with the frequency changes and thus would not be able to divert the torpedo from its target.

Spread Spectrum

Lamarr's idea was ahead of its time. The digital technology that allowed efficient operation of her system wasn't invented until decades later. However, after 1962, frequency hopping was adopted and used in U.S. military communications. It was the development of wireless phones, however, that benefited the most from Lamarr's concept.

Cordless phones and other wireless technologies operate by breaking their signals into smaller parts, called packets. The frequency of the packets switches rapidly, preventing interference with other calls and enabling millions of callers to use the same narrow band of frequencies.

A torpedo is launched during World War II.

Science online

For more information, visit booko.msscience.com/time

Reviewing Main Ideas

Section 1 ### The Nature of Electromagnetic Waves

1. Vibrating charges generate vibrating electric and magnetic fields. These vibrating fields travel through space and are called electromagnetic waves.

2. Electromagnetic waves have wavelength, frequency, amplitude, and carry energy.

Section 2 ### The Electromagnetic Spectrum

1. Radio waves have the longest wavelength and lowest energy. Radar uses radio waves to locate objects.

2. All objects emit infrared waves. Most objects you see reflect the visible light emitted by a source of light.

3. Ultraviolet waves have a higher frequency and carry more energy than visible light.

4. X rays and gamma rays are highly penetrating and can be dangerous to living organisms.

Section 3 ### Using Electromagnetic Waves

1. Communications systems use electromagnetic waves to transmit information.

2. Radio and TV stations use modulated carrier waves to transmit information.

3. Cordless and cell phones use radio waves to communicate between the mobile phone and a base station.

4. Radio waves are used to send information between communications satellites and ground stations on Earth.

Visualizing Main Ideas

Copy and complete the following spider map about electromagnetic waves.

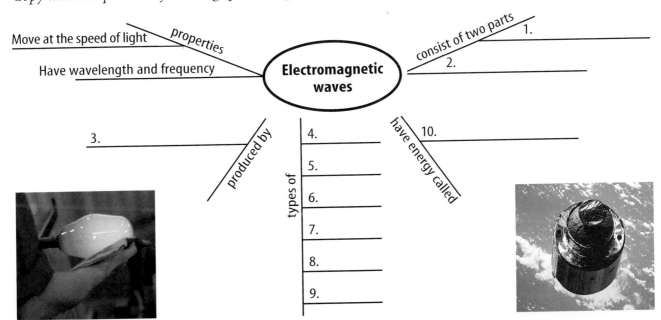

Using Vocabulary

carrier wave p. 82
electromagnetic spectrum
  p. 71
electromagnetic wave
  p. 66
gamma ray p. 76
Global Positioning System
  p. 85

infrared wave p. 73
radiant energy p. 70
radio wave p. 72
ultraviolet radiation p. 75
visible light p. 74
X ray p. 76

Explain the difference between the terms in each of the following pairs.

1. infrared wave—radio wave

2. radio wave—carrier wave

3. communications satellite—Global Positioning System

4. visible light—ultraviolet radiation

5. X ray, gamma ray

6. electromagnetic wave—radiant energy

7. carrier wave—AM radio signal

8. infrared wave—ultraviolet wave

Checking Concepts

Choose the word or phrase that best answers the question.

9. Which of the following transformations can occur in a radio antenna?
   A) radio waves to sound waves
   B) radio waves to an electric signal
   C) radio waves to infrared waves
   D) sound waves to radio waves

10. Electromagnetic waves with wavelengths between about 0.7 millionths of a meter and 0.4 millionths of a meter are which of the following?
    A) gamma rays
    B) microwaves
    C) radio waves
    D) visible light

11. Which of the following is the speed of light in space?
    A) 186,000 km/s
    B) 300,000 km/s
    C) 3,000,000 km/s
    D) 30,000 km/s

12. Which of the following types of electro-magnetic waves has the lowest frequency?
    A) infrared waves
    B) visible light
    C) radio waves
    D) gamma rays

13. Compared to an electric charge that is not moving, a moving electric charge is sur-rounded by which of the following addi-tional fields?
    A) magnetic
    B) microwave
    C) electric
    D) gravitational

14. Most of the electromagnetic waves emit-ted by an object at room temperature are which of the following?
    A) visible light
    B) radio waves
    C) infrared waves
    D) X rays

15. Which of the following color of visible light has the highest frequency?
    A) green
    B) blue
    C) yellow
    D) red

16. Which type of electromagnetic waves are completely absorbed by Earth's atmosphere?
    A) radio waves
    B) infrared waves
    C) gamma rays
    D) visible light

17. Sunburn is caused by excessive exposure to which of the following?
    A) ultraviolet waves
    B) infrared waves
    C) visible light
    D) gamma rays

18. How does the frequency of a gamma ray change as its wavelength decreases?
    A) It increases.
    B) It decreases.
    C) It doesn't change.
    D) The frequency depends on the speed.

Thinking Critically

19. Infer why communications systems usually use radio waves to transmit information.

20. Classify List the colors of the visible light spectrum in order of increasing frequency.

21. Compare and contrast an electromagnetic wave with a transverse wave traveling along a rope.

22. Explain Some stars form black holes when they collapse. These black holes sometimes can be found by detecting X rays and gamma rays that are emitted as matter falls into the black hole. Explain why it would be difficult to detect these X rays and gamma rays using detectors at Earth's surface.

Use the table below to answer question 23.

Speed of Light in Various Materials	
Materials	**Speed (km/s)**
Air	300,000
Water	226,000
Polystyrene Plastic	189,000
Diamond	124,000

23. Calculate A radio wave has a frequency of 500,000 Hz. If the radio wave has the same frequency in air as in water, what is the ratio of the wavelength of the radio wave in air to its wavelength in water?

24. Explain how you could determine if there are electromagnetic waves traveling in a closed, completely dark room in a building.

25. Infer Light waves from a distant galaxy take 300 million years to reach Earth. How does the age of the galaxy when it emitted the light waves compare with the age of the galaxy when we see the light waves?

26. Concept Map Electromagnetic waves are grouped according to their frequencies. In the following concept map, write each frequency group and one way humans make use of the electromagnetic waves in that group. For example, in the second set of ovals, you might write *X rays* and *to see inside the body*. **Do not write in this book.**

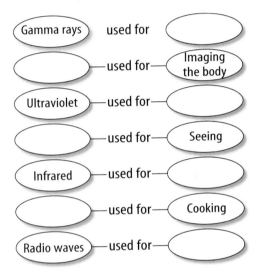

Performance Activities

27. Oral Presentation Explain to the class how a radio signal is generated, transmitted, and received.

28. Poster Make a poster showing the parts of the electromagnetic spectrum. Show how frequency, wavelength, and energy change throughout the spectrum. How is each wave generated? What are some uses of each?

Applying Math

29. Distance How long would it take a radio signal to travel from Earth to the Moon, a distance of 384,000 km?

30. Wavelength The frequency of a popular AM radio station is 720 kHz. What is the wavelength of the radio waves broadcast by this station?

Part 1 | Multiple Choice

Record your answers on the answer sheet provided by your teacher or on a sheet of paper.

1. Which of the following types of electromagnetic waves has a frequency greater than visible light?
   A. infrared waves
   B. radio waves
   C. ultraviolet waves
   D. microwaves

2. Which of the following properties of a transverse wave is the distance from one crest to the next?
   A. intensity
   B. amplitude
   C. frequency
   D. wavelength

3. Which of the following types of electromagnetic waves enables your body to produce vitamin D?
   A. gamma rays
   B. ultraviolet waves
   C. visible light
   D. infrared waves

Use the illustration below to answer question 4.

Electromagnetic Waves from the Sun

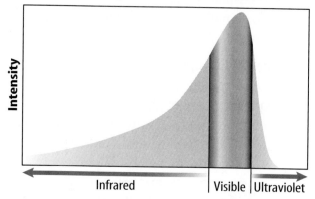

4. How does the intensity of ultraviolet waves emitted by the Sun change as the wavelength of the ultraviolet waves decreases?
   A. The intensity increases.
   B. The intensity decreases.
   C. The intensity doesn't change.
   D. The intensity increases, then decreases.

5. The color of visible light waves depends on which of the following wave properties?
   A. wavelength
   B. amplitude
   C. direction
   D. speed

6. Which of the following is NOT true about electromagnetic waves?
   A. They can travel through matter.
   B. They move by transferring matter.
   C. They are produced by vibrating charges.
   D. They can travel through empty space.

Use the illustration below to answer question 7.

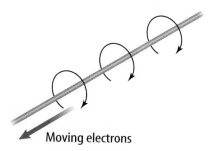
Moving electrons

7. Which of the following is represented by the circular lines around the current-carrying wire?
   A. direction of current
   B. electric and magnetic field lines
   C. magnetic field lines
   D. electric field lines

8. How are gamma rays produced?
   A. by vibrating electric fields
   B. by vibrating magnetic fields
   C. by the absorption of infrared waves
   D. by nuclear fission or fusion

9. Earth's atmosphere is transparent to which type of electromagnetic waves?
   A. gamma rays
   B. ultraviolet waves
   C. infrared waves
   D. radio waves

Part 2 | Short Response/Grid In

Record your answers on the answer sheet provided by your teacher or on a sheet of paper.

Use the photograph below to answer question 10.

10. If the microwaves produced in a microwave oven have a frequency of 2,450 MHz, what is the wavelength of the microwaves?

11. You turn on a lamp that is plugged into an electric outlet. Does a magnetic field surround the wire that connects the lamp to the outlet? Explain.

12. A carrier wave broadcast by a radio station has a wavelength of 3.0 m. What is the frequency of the carrier wave?

13. Explain how the wavelengths of the electromagnetic waves emitted by an object change as the temperature of the object increases.

14. Explain why X rays can form images of dense tissues in the human body.

15. If the planet Mars is 80,000,000 km from Earth, how long will it take an electromagnetic wave to travel from Earth to Mars?

Test-Taking Tip

Recheck Answers Double check your answers before turning in the test.

Part 3 | Open Ended

Record your answers on a sheet of paper.

16. Describe the sequence of events that occur when a radar station detects an airplane and determines the distance to the plane.

17. Explain why infrared detectors on satellites can detect regions covered by vegetation.

18. The carrier waves broadcast by a radio station are altered in order to transmit information. The two ways of altering a carrier wave are amplitude modulation (AM) and frequency modulation (FM). Draw a carrier wave, an AM wave, and an FM wave.

19. Describe the effect of the ozone layer on electromagnetic waves that strike Earth's atmosphere.

20. List the energy conversions that occur when a song recorded on a CD is broadcast as radio waves and then reproduced as sound.

Use the illustration below to answer questions 21 and 22.

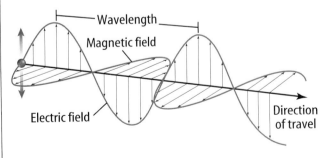

21. Explain how the vibrating electric and magnetic fields are produced.

22. Infer how the directions of the electric field and the magnetic field are related to the direction that the electromagnetic wave travels.

Light, Mirrors, and Lenses

The BIG Idea

Light waves can be absorbed, reflected, and transmitted by matter.

SECTION 1
Properties of Light
Main Idea A source of light gives off light rays that travel outward in all directions.

SECTION 2
Reflection and Mirrors
Main Idea When a light ray is reflected from a surface, the angle of incidence equals the angle of reflection.

SECTION 3
Refraction and Lenses
Main Idea A light ray changes direction when it moves from one material into another and changes speed.

SECTION 4
Using Mirrors and Lenses
Main Idea Lenses and mirrors are used to form images of objects that cannot be seen with the human eye.

Seeing the Light

This lighthouse at Pigeon Point, California, produces beams of light that can be seen for many miles. These intense light beams are formed in the same way as a flashlight beam. The key ingredient is a curved mirror that reflects the light from a bright source.

Science Journal Describe how you use mirrors and lenses during a typical day.

Start-Up Activities

Bending Light

Everything you see results from light waves entering your eyes. These light waves are either given off by objects, such as the Sun and lightbulbs, or reflected by objects, such as trees, books, and people. Lenses and mirrors can cause light to change direction and make objects seem larger or smaller.

1. Place two paper cups next to each other and put a penny in the bottom of each cup.

2. Fill one of the cups with water and observe how the penny looks.

3. Looking straight down at the cups, slide the cup with no water away from you just until you can no longer see the penny.

4. Pour water into this cup and observe what seems to happen to the penny.

5. **Think Critically** In your Science Journal, record your observations. Did adding water make the cup look deeper or shallower?

Preview this chapter's content and activities at
booko.msscience.com

Light, Mirrors, and Lenses
Make the following Foldable to help you understand the properties of and the relationship between light, mirrors, and lenses.

STEP 1 **Fold** a sheet of pape in half lengthwise. Make the back edge about 5 cm longer than the front edge.

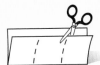

STEP 2 **Turn** the paper so the fold is on the bottom. Then **fold** it into thirds.

STEP 3 **Unfold and cut** only the top layer along folds to make three tabs.

STEP 4 **Label** the Foldable as shown.

Summarize in a Table As you read the chapter, summarize the information you find about light, mirrors, lenses.

0 ◆ 95

Get Ready to Read

Questioning

① Learn It! Asking questions helps you to understand what you read. As you read, think about the questions you'd like answered. Often you can find the answer in the next paragraph or section. Learn to ask good questions by asking who, what, when, where, why, and how.

② Practice It! Read the following passage from Section 3.

> Optical fibers are used most commonly in the communications industry. For example, television programs, computer information, and phone conversations can be coded into light signals. These signals can then be sent from one place to another using optical fibers. Because of total internal reflection, signals cannot leak from one fiber to another, causing interference. As a result, the signal is transmitted clearly.
>
> — *from page 112*

Here are some questions you might ask about this paragraph:

- How are optical fibers used by the communications industry?
- What type of signals are sent through optical fibers?
- Why are signals transmitted clearly in optical fibers?

③ Apply It! As you read the chapter, look for answers to lesson headings that are in the form of questions.

Reading Tip

Test yourself. Create questions and then read to find answers to your questions.

Target Your Reading

Use this to focus on the main ideas as you read the chapter.

① **Before you read** the chapter, respond to the statements below on your worksheet or on a numbered sheet of paper.
- Write an **A** if you **agree** with the statement.
- Write a **D** if you **disagree** with the statement.

② **After you read** the chapter, look back to this page to see if you've changed your mind about any of the statements.
- If any of your answers changed, explain why.
- Change any false statements into true statements.
- Use your revised statements as a study guide.

Science Online
Print out a worksheet of this page at booko.msscience.com

Before You Read A or D		Statement	After You Read A or D
	1	All objects give off light on their own.	
	2	You see an object when light rays travel from your eyes to the object.	
	3	The color of an object depends on the wavelengths of the light waves reflected from the object.	
	4	Light rays obey the law of reflection only if the reflecting surface is very smooth.	
	5	Light waves travel at the same speed in all materials.	
	6	A lens causes all light rays to pass through the focal point of the lens.	
	7	The image formed by a lens depends on how far the object is from the lens.	
	8	The purpose of the large concave mirror in a reflecting telescope is to magnify objects.	
	9	A laser beam contains a single wavelength of light.	

Properties of Light

What You'll Learn
- **Describe** the wave nature of light.
- **Explain** how light interacts with materials.
- **Determine** why objects appear to have color.

Why It's Important
Everything you see comes from information carried by light waves.

⟲ Review Vocabulary
electromagnetic waves: waves created by vibrating electric charges that can travel through space or through matter

New Vocabulary
- light ray
- medium

What is light?

Drop a rock on the smooth surface of a pond and you'll see ripples spread outward from the spot where the rock struck. The rock produced a wave much like the one in **Figure 1.** A wave is a disturbance that carries energy through matter or space. The matter in this case is the water, and the energy originally comes from the impact of the rock. As the ripples spread out, they carry some of that energy.

Light is another type of wave that carries energy. A source of light such as the Sun or a lightbulb gives off light waves into space, just as the rock hitting the pond causes waves to form in the water. But while the water waves spread out only on the surface of the pond, light waves spread out in all directions from the light source. **Figure 1** shows how light waves travel.

Sometimes, however, it is easier to think of light in a different way. A **light ray** is a narrow beam of light that travels in a straight line. You can think of a source of light as giving off, or emitting, a countless number of light rays that are traveling away from the source in all directions.

Figure 1 Light moves away in all directions from a light source, just as ripples spread out on the surface of water.

A source of light, such as a lightbulb, gives off light rays that travel away from the light source in all directions.

Ripples on the surface of a pond are produced by an object hitting the water. The ripples spread out from the point of impact.

Light Travels Through Space There is, however, one important difference between light waves and the water wave ripples on a pond. If the pond dried up and had no water, ripples could not form. Waves on a pond need a material—water—in which to travel. The material through which a wave travels is called a **medium.** Light is an electromagnetic wave and doesn't need a medium in which to travel. Electromagnetic waves can travel in a vacuum, as well as through materials such as air, water, and glass.

Light and Matter

What can you see when you are in a closed room with no windows and the lights out? You can see nothing until you turn on a light or open a door to let in light from outside the room. Most objects around you do not give off light on their own. They can be seen only if light waves from another source bounce off them and into your eyes, as shown in **Figure 2.** The process of light striking an object and bouncing off is called reflection. Right now, you can see these words because light emitted by a source of light is reflecting from the page and into your eyes. Not all the light rays reflected from the page strike your eyes. Light rays striking the page are reflected in many directions, and only some of these rays enter your eyes.

Reading Check *What must happen for you to see most objects?*

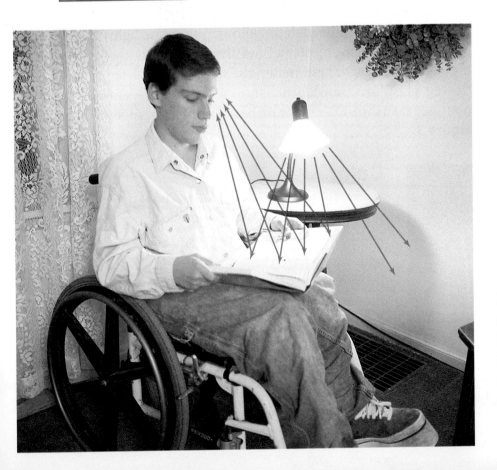

Mini LAB

Observing Colors in the Dark

Procedure
1. Get six pieces of **paper** that are different colors and about 10 cm × 10 cm.
2. Darken a room and wait 10 min for your eyes to adjust to the darkness.
3. Write on each paper what color you think the paper is.
4. Turn on the lights and see if your night vision correctly detected the colors.

Analysis
1. If the room were perfectly dark, what would you see? Explain.
2. Your eyes contain rod cells and cone cells. Rod cells enable you to see in dim light, but don't detect color. Cone cells enable you to see color, but do not work in dim light. Which type of cell was working in the darkened room? Explain.

Try at Home

Figure 2 Light waves are given off by the lightbulb. Some of these light waves hit the page and are reflected. The student sees the page when some of these reflected waves enter the student's eyes.

An opaque object allows no light to pass through it.

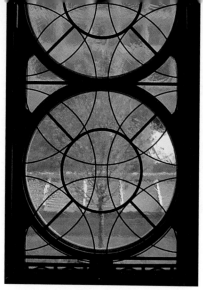

A translucent object allows some light to pass through it.

A transparent object allows almost all light to pass through it.

Figure 3 Materials are opaque, translucent, or transparent, depending on how much light passes through them.
Infer *which type of material reflects the least amount of light.*

Figure 4 A beam of white light passing through a prism is separated into many colors.
Describe *the colors you see emerging from the prism.*

Opaque, Translucent, and Transparent When light waves strike an object, some of the waves are absorbed by the object, some are reflected by it, and some might pass through it. What happens to light when it strikes the object depends on the material that the object is made of.

All objects reflect and absorb some light waves. Materials that let no light pass through them are opaque (oh PAYK). You cannot see other objects through opaque materials. On the other hand, you clearly can see other objects through materials such as glass and clear plastic that allow nearly all the light that strikes them to pass through. These materials are transparent. A third type of material allows only some light to pass through. Although objects behind these materials are visible, they are not clear. These materials, such as waxed paper and frosted glass, are translucent (trans LEW sent). Examples of opaque, translucent, and transparent objects are shown in **Figure 3.**

Color

The light from the Sun might look white, but it is a mixture of colors. Each different color of light is a light wave with a different wavelength. Red light waves have the longest wavelengths and violet light waves have the shortest wavelengths. As shown in **Figure 4,** white light is separated into different colors when it passes through a prism. The colors in white light range from red to violet. When light waves from all these colors enter the eye at the same time, the brain interprets the mixture as being white.

A pair of gym shoes and socks as seen under white light.

The same shoes and socks photographed through a red filter.

Why do objects have color? Why does grass look green or a rose look red? When a mixture of light waves strikes an object that is not transparent, the object absorbs some of the light waves. Some of the light waves that are not absorbed are reflected. If an object reflects red waves and absorbs all the other waves, it looks red. Similarly, if an object looks blue, it reflects only blue light waves and absorbs all the others. An object that reflects all the light waves that strike it looks white, while one that reflects none of the light waves that strike it looks black. **Figure 5** shows gym shoes and socks as seen under white light and as seen when viewed through a red filter that allows only red light to pass through it.

Primary Light Colors How many colors exist? People often say white light is made up of red, orange, yellow, green, blue, and violet light. This isn't completely true, though. Many more colors than this exist. In reality, most humans can distinguish thousands of colors, including some such as brown, pink, and purple, that are not found among the colors of the rainbow.

Light of almost any color can be made by mixing different amounts of red, green, and blue light. Red, green, and blue are known as the primary colors. Look at **Figure 6.** White light is produced where beams of red, green, and blue light overlap. Yellow light is produced where red and green light overlap. You see the color yellow because of the way your brain interprets the combination of the red and green light striking your eye. This combination of light waves looks the same as yellow light produced by a prism, even though these light waves have only a single wavelength.

Figure 5 The color of an object depends on the light waves it reflects.
Infer *why the blue socks look black when viewed under red light.*

Figure 6 By mixing light from the three primary colors—red, blue, and green—almost all of the visible colors can be made.

Figure 7 The three primary color pigments—yellow, magenta, and cyan—can form almost all the visible colors when mixed together in various amounts.

Primary Pigment Colors Materials like paint that are used to change the color of other objects, such as the walls of a room or an artist's canvas, are called pigments. Mixing pigments together forms colors in a different way than mixing colored lights does.

Like all materials that appear to be colored, pigments absorb some light waves and reflect others. The color of the pigment you see is the color of the light waves that are reflected from it. However, the primary pigment colors are not red, blue, and green—they are yellow, magenta, and cyan. You can make almost any color by mixing different amounts of these primary pigment colors, as shown in **Figure 7.**

Although primary pigment colors are not the same as the primary light colors, they are related. Each primary pigment color results when a pigment absorbs a primary light color. For example, a yellow pigment absorbs blue light and it reflects red and green light, which you see as yellow. A magenta pigment, on the other hand, absorbs green light and reflects red and blue light, which you see as magenta. Each of the primary pigment colors is the same color as white light with one primary color removed.

section 1 review

Summary

Light and Matter

- Light is an electromagnetic wave that can travel in a vacuum as well as through matter.
- When light waves strike an object some light waves might be absorbed by the object, some waves might be reflected from the object, and some waves might pass through the object.
- Materials can be opaque, translucent, or transparent, depending on how much light passes through the material.

Color

- Light waves with different wavelengths have different colors.
- White light is a combination of all the colors ranging from red to violet.
- The color of an object is the color of the light waves that it reflects.
- The primary light colors are red, green, and blue. The primary pigment colors are yellow, magenta and cyan.

Self Check

1. **Diagram** the path followed by a light ray that enters one of your eyes when you are reading at night in a room.

2. **Determine** the colors that are reflected from an object that appears black.

3. **Compare and contrast** primary light colors and primary pigment colors.

4. **Describe** the difference between an opaque object and a transparent object.

5. **Think Critically** A white shirt is viewed through a filter that allows only blue light to pass through the filter. What color will the shirt appear to be?

Applying Skills

6. **Draw Conclusions** A black plastic bowl and a white plastic bowl are placed in sunlight. After 15 minutes, the temperature of the black bowl is higher than the temperature of the white bowl. Which bowl absorbs more light waves and which bowl reflects more light waves?

Science Online booko.msscience.com/self_check_quiz

Reflection and Mirrors

The Law of Reflection

You've probably noticed your image on the surface of a pool or lake. If the surface of the water was smooth, you could see your face clearly. If the surface of the water was wavy, however, your face might have seemed distorted. The image you saw was the result of light reflecting from the surface and traveling to your eyes. How the light was reflected determined the sharpness of the image you saw.

When a light ray strikes a surface and is reflected, as in **Figure 8,** the reflected ray obeys the law of reflection. Imagine a line that is drawn perpendicular to the surface where the light ray strikes. This line is called the normal to the surface. The incoming ray and the normal form an angle called the angle of incidence. The reflected light ray forms an angle with the normal called the angle of reflection. According to the **law of reflection,** the angle of incidence is equal to the angle of reflection. This is true for any surface, no matter what material it is made of.

Reflection from Surfaces

Why can you see your reflection in some surfaces and not others? Why does a piece of shiny metal make a good mirror, but a piece of paper does not? The answers have to do with the smoothness of each surface.

as you read

***What* You'll Learn**

- **Explain** how light is reflected from rough and smooth surfaces.
- **Determine** how mirrors form an image.
- **Describe** how concave and convex mirrors form an image.

***Why* It's Important**

Mirrors can change the direction of light waves and enable you to see images, such as your own face.

Review Vocabulary

normal: a line drawn perpendicular to a surface or line

New Vocabulary

- law of reflection
- focal point
- focal length

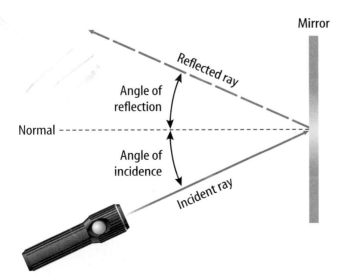

Figure 8 A light ray strikes a surface and is reflected. The angle of incidence is always equal to the angle of reflection. This is the law of reflection.

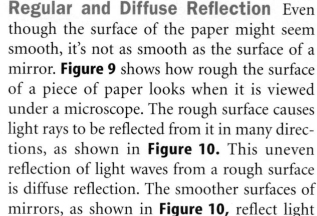

Figure 9 A highly magnified view of the surface of a sheet of paper shows that the paper is made of many cellulose wood fibers that make the surface rough and uneven.

Magnification: 80×

Regular and Diffuse Reflection Even though the surface of the paper might seem smooth, it's not as smooth as the surface of a mirror. **Figure 9** shows how rough the surface of a piece of paper looks when it is viewed under a microscope. The rough surface causes light rays to be reflected from it in many directions, as shown in **Figure 10.** This uneven reflection of light waves from a rough surface is diffuse reflection. The smoother surfaces of mirrors, as shown in **Figure 10,** reflect light waves in a much more regular way. For example, parallel rays remain parallel after they are reflected from a mirror. Reflection from mirrors is known as regular reflection. Light waves that are regularly reflected from a surface form the image you see in a mirror or any other smooth surface. Whether a surface is smooth or rough, every light ray that strikes it obeys the law of reflection.

✅ **Reading Check** *Why does a rough surface cause a diffuse reflection?*

Scattering of Light When diffuse reflection occurs, light waves that were traveling in a single direction are reflected and then travel in many different directions. Scattering occurs when light waves traveling in one direction are made to travel in many different directions. Scattering also can occur when light waves strike small particles, such as dust. You may have seen dust particles floating in a beam of sunlight. When the light waves in the sunbeam strike a dust particle, they are scattered in all directions. You see the dust particles as bright specks of light when some of these scattered light waves enter your eye.

Figure 10 The roughness of a surface determines whether it looks like a mirror.

A rough surface causes parallel light rays to be reflected in many different directions.

A smooth surface causes parallel light rays to be reflected in a single direction. This type of surface looks like a mirror.

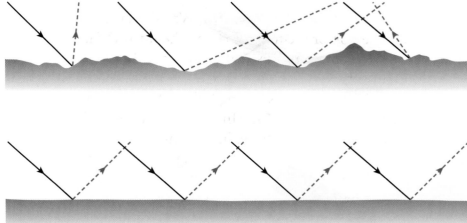

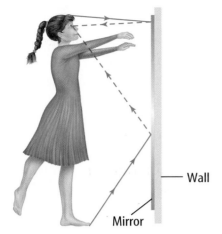

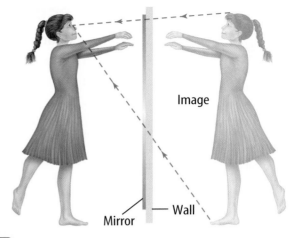

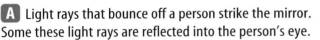

A Light rays that bounce off a person strike the mirror. Some these light rays are reflected into the person's eye.

B The light rays that are shown entering the person's eye seem to be coming from a person behind the mirror.

Reflection by Plane Mirrors Did you glance in the mirror before leaving for school this morning? If you did, you probably looked at your reflection in a plane mirror. A plane mirror is a mirror with a flat reflecting surface. In a plane mirror, your image looks much the same as it would in a photograph. However, you and your image are facing in opposite directions. This causes your left side and your right side to switch places on your mirror image. Also, your image seems to be coming from behind the mirror. How does a plane mirror form an image?

Reading Check *What is a plane mirror?*

Figure 11 shows a person looking into a plane mirror. Light waves from the Sun or another source of light strike each part of the person. These light rays bounce off the person according to the law of reflection, and some of them strike the mirror. The rays that strike the mirror also are reflected according to the law of reflection. **Figure 11A** shows the path traveled by a few of the rays that have been reflected off the person and reflected back to the person's eye by the mirror.

The Image in a Plane Mirror Why does the image you see in a plane mirror seem to be behind the mirror? This is a result of how your brain processes the light rays that enter your eyes. Although the light rays bounced off the mirror's surface, your brain interprets them as having followed the path shown by the dashed lines in **Figure 11B.** In other words, your brain always assumes that light rays travel in straight lines without changing direction. This makes the reflected light rays look as if they are coming from behind the mirror, even though no source of light is there. The image also seems to be the same distance behind the mirror as the person is in front of the mirror.

Figure 11 A plane mirror forms an image by changing the direction of light rays.
Describe *how you and your image in a plane mirror are different.*

Light Waves and Photons When an object like a marble or a basketball bounces off a surface, it obeys the law of reflection. Because light also obeys the law of reflection, people once thought that light must be a stream of particles. Today, experiments have shown that light can behave as though it were both a wave and a stream of energy bundles called photons. Read an article about photons and write a description in your Science Journal.

Concave and Convex Mirrors

Some mirrors are not flat. A concave mirror has a surface that is curved inward, like the bowl of a spoon. Unlike plane mirrors, concave mirrors cause light rays to come together, or converge. A convex mirror, on the other hand, has a surface that curves outward, like the back of a spoon. Convex mirrors cause light waves to spread out, or diverge. These two types of mirrors form images that are different from the images that are formed by plane mirrors. Examples of a concave and a convex mirror are shown in **Figure 12.**

 **Reading Check** *What's the difference between a concave and convex mirror?*

Concave Mirrors The way in which a concave mirror forms an image is shown in **Figure 13.** A straight line drawn perpendicular to the center of a concave or convex mirror is called the optical axis. Light rays that travel parallel to the optical axis and strike the mirror are reflected so that they pass through a single point on the optical axis called the **focal point.** The distance along the optical axis from the center of the mirror to the focal point is called the **focal length.**

The image formed by a concave mirror depends on the position of the object relative to its focal point. If the object is farther from the mirror than the focal point, the image appears to be upside down, or inverted. The size of the image decreases as the object is moved farther away from the mirror. If the object is closer to the mirror than one focal length, the image is upright and gets smaller as the object moves closer to the mirror.

A concave mirror can produce a focused beam of light if a source of light is placed at the mirror's focal point, as shown in **Figure 13.** Flashlights and automobile headlights use concave mirrors to produce directed beams of light.

Figure 12 Convex and concave mirrors have curved surfaces.

A concave mirror has a surface that's curved inward.

A convex mirror has a surface that's curved outward.

Figure 13

Glance into a flat plane mirror and you'll see an upright image of yourself. But look into a concave mirror, and you might see yourself larger than life, right side up, or upside down—or not at all! This is because the way a concave mirror forms an image depends on the position of an object in front of the mirror, as shown here.

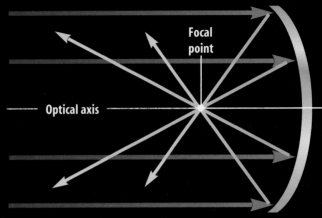

A concave mirror reflects all light rays traveling parallel to the optical axis so that they pass through the focal point.

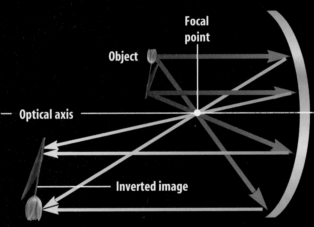

When an object, such as this flower, is placed beyond the focal point, the mirror forms an image that is inverted.

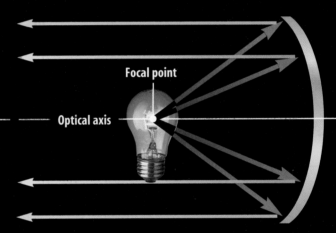

When a source of light is placed at the focal point, a beam of parallel light rays is formed. The concave mirror in a flashlight, for example, creates a beam of parallel light rays.

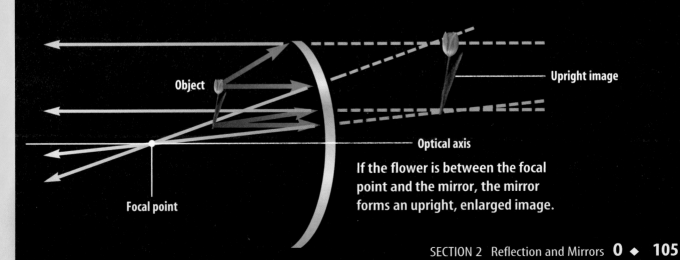

If the flower is between the focal point and the mirror, the mirror forms an upright, enlarged image.

Figure 14 A convex mirror is a mirror that curves outward.

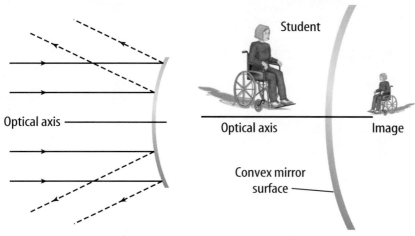

Student

Optical axis

Optical axis

Image

Convex mirror surface

A convex mirror causes light rays that are traveling parallel to the optical axis to spread apart after they are reflected.

No matter how far an object is from a convex mirror, the image is always upright and smaller than the object.

Convex Mirrors A convex mirror has a reflecting surface that curves outward and causes light rays to spread apart, or diverge, as shown in **Figure 14**. Like the image formed by plane mirror, the image formed by a convex mirror seems to be behind the mirror. **Figure 14** shows that the image always is upright and smaller than the object.

Convex mirrors often are used as security mirrors in stores and as outside rearview mirrors on cars and other vehicles. You can see a larger area reflected in a convex mirror than in other mirrors.

section ② review

Summary

Reflection and Plane Mirrors

- The law of reflection states that the angle of incidence equals the angle of reflection.
- A regular reflection is produced by a smooth surface, such as a mirror. A rough surface forms a diffuse reflection.
- Scattering occurs when light rays traveling in one direction are made to travel in many directions.
- A plane mirror forms a image that is reversed left to right and seems to be behind the mirror.

Concave and Convex Mirrors

- Concave mirrors curve inward and make light rays converge.
- Images formed by a concave mirror can be either upright or inverted and can vary from larger to smaller than the object.
- Convex mirrors curve outward and make light rays diverge.
- Images formed by a convex mirror are always upright and smaller than the object.

Self Check

1. **Describe** the image formed by a concave mirror when an object is less than one focal length from the mirror.
2. **Explain** why concave mirrors are used in flashlights and automobile headlights.
3. **Describe** If an object is more than one focal length from a concave mirror, how does the image formed by the mirror change as the object moves farther from the mirror?
4. **Determine** which light rays striking a concave mirror are reflected so that they pass through the focal point.
5. **Think Critically** After you wash and wax a car, you can see your reflection in the car's surface. Before you washed and waxed the car, no reflection could be seen. Explain.

Applying Skills

6. **Use a Spreadsheet** Make a table using a spreadsheet comparing the images formed by plane, concave, and convex mirrors. Include in your table how the images depend on the distance of the object from the mirror.

Science Online booko.msscience.com/self_check_quiz

Reflection from a Plane Mirror

A light ray strikes the surface of a plane mirror and is reflected. Does a relationship exist between the direction of the incoming light ray and the direction of the reflected light ray?

◉ Real-World Question

How does the angle of incidence compare with the angle of reflection for a plane mirror?

Goals
- ■ **Measure** the angle of incidence and the angle of reflection for a light ray reflected from a plane mirror.

Materials
flashlight
protractor
metric ruler
scissors
tape

small plane mirror, at least 10 cm on a side
black construction paper
modeling clay
white unlined paper

Safety Precautions

◉ Procedure

1. With the scissors, cut a slit in the construction paper and tape it over the flashlight lens.

2. Place the mirror at one end of the unlined paper. Push the mirror into lumps of clay so it stands vertically, and tilt the mirror so it leans slightly toward the table.

3. **Measure** with the ruler to find the center of the bottom edge of the mirror, and mark it. Then use the protractor and the ruler to draw a line on the paper perpendicular to the mirror from the mark. Label this line P.

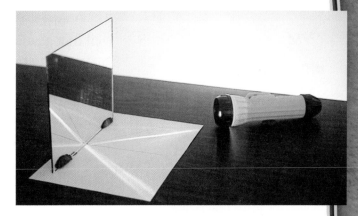

4. Draw lines on the paper from the center mark at angles of 30°, 45°, and 60° to line P.

5. Turn on the flashlight and place it so the beam is along the 60° line. This is the angle of incidence. Measure and record the angle that the reflected beam makes with line P. This is the angle of reflection. If you cannot see the reflected beam, slightly increase the tilt of the mirror.

6. Repeat step 5 for the 30°, 45°, and P lines.

◉ Conclude and Apply

Infer from your results the relationship between the angle of incidence and the angle of reflection.

*C*ommunicating
Your Data

Make a poster that shows your measured angles of reflection for angles of incidence of 30°, 45°, and 60°. Write the relationship between the angles of incidence and reflection at the bottom.

Refraction and Lenses

What **You'll Learn**

- **Determine** why light rays refract.
- **Explain** how convex and concave lenses form images.

Why **It's Important**

Many of the images you see every day in photographs, on TV, and in movies are made using lenses.

Review Vocabulary

refraction: bending of a wave as it changes speed, moving from one medium to another

New Vocabulary
- lens
- convex lens
- concave lens

Bending of Light Rays

Objects that are in water can sometimes look strange. A pencil in a glass of water sometimes looks as if it's bent, or as if the part of the pencil in air is shifted compared to the part in water. A penny that can't be seen at the bottom of a cup suddenly appears as you add water to the cup. Illusions such as these are due to the bending of light rays as they pass from one material to another. What causes light rays to change direction?

The Speeds of Light The speed of light in empty space is about 300 million m/s. Light passing through a material such as air, water, or glass, however, travels more slowly than this. This is because the atoms that make up the material interact with the light waves and slow them down. **Figure 15** compares the speed of light in some different materials.

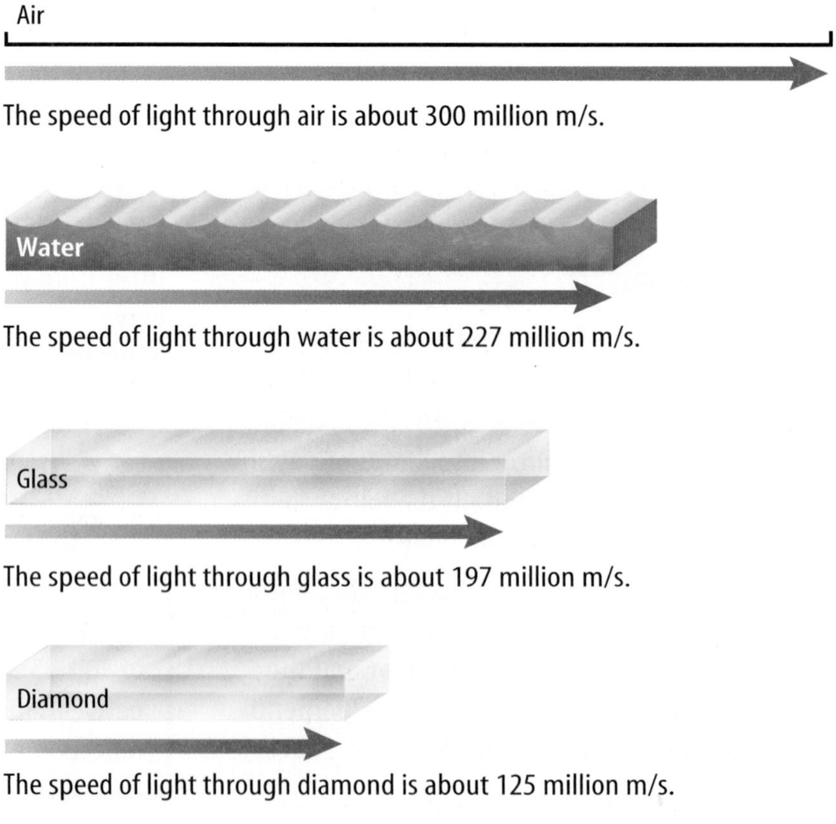

Air

The speed of light through air is about 300 million m/s.

Water

The speed of light through water is about 227 million m/s.

Glass

The speed of light through glass is about 197 million m/s.

Diamond

The speed of light through diamond is about 125 million m/s.

Figure 15 Light travels at different speeds in different materials.

The Refraction of Light Waves

Light rays from the part of a pencil that is underwater travel through water, glass, and then air before they reach your eye. The speed of light is different in each of these mediums. What happens when a light wave travels from one medium into another in which its speed is different? If the wave is traveling at an angle to the boundary between the two media, it changes direction, or bends. This bending is due to the change in speed the light wave undergoes as it moves from one medium into the other. The bending of light waves due to a change in speed is called refraction. **Figure 16** shows an example of refraction. The greater the change in speed is, the more the light wave bends, or refracts.

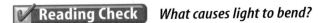

Reading Check *What causes light to bend?*

Why does a change in speed cause the light wave to bend? Think about what happens to the wheels of a car as they move from pavement to mud at an angle, as in **Figure 17.** The wheels slip a little in the mud and don't move forward as fast as they do on the pavement. The wheel that enters the mud first gets slowed down a little, but the other wheel on that axle continues at the original speed. The difference in speed between the two wheels then causes the wheel axle to turn, so the car turns a little. Light waves behave in the same way.

Imagine again a light wave traveling at an angle from air into water. The first part of the wave to enter the water is slowed, just as the car wheel that first hit the mud was slowed. The rest of the wave keeps slowing down as it moves from the air into the water. As long as one part of the light wave is moving faster than the rest of the wave, the wave continues to bend.

Figure 16 A light ray is bent as it slows down traveling from air into water.

Figure 17 An axle turns as the wheels cross the boundary between pavement and mud. **Predict** *how the axle would turn if the wheels were going from mud to pavement.*

Convex and Concave Lenses

Do you like photographing your friends and family? Have you ever watched a bird through binoculars or peered at something tiny through a magnifying glass? All of these activities involve the use of lenses. A **lens** is a transparent object with at least one curved side that causes light to bend. The amount of bending can be controlled by making the sides of the lenses more or less curved. The more curved the sides of a lens are, the more light will be bent after it enters the lens.

Figure 18 A convex lens forms an image that depends on the distance from the object to the lens.

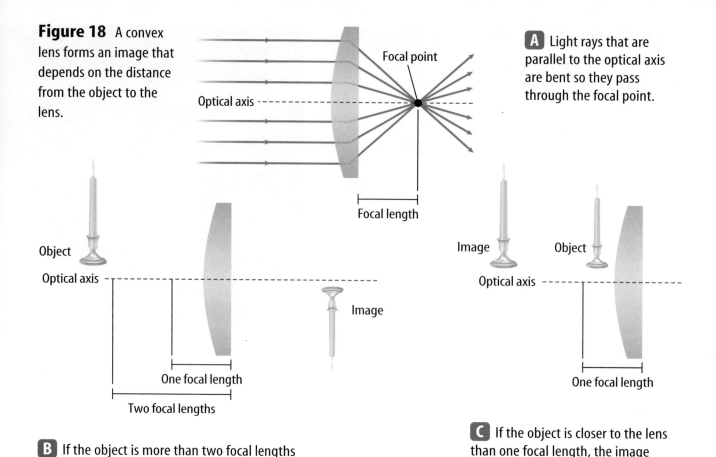

Focal point

Optical axis

Focal length

A Light rays that are parallel to the optical axis are bent so they pass through the focal point.

Object

Optical axis

Image

One focal length

Two focal lengths

B If the object is more than two focal lengths from the lens, the image formed is smaller than the object and inverted.

Image Object

Optical axis

One focal length

C If the object is closer to the lens than one focal length, the image formed is enlarged and upright.

Convex Lenses A lens that is thicker in the center than at the edges is a **convex lens**. In a convex lens, light rays traveling parallel to the optical axis are bent so they pass through the focal point, as shown in **Figure 18A.** The more curved the lens is, the closer the focal point is to the lens, and so the shorter the focal length of the lens is. Because convex lenses cause light waves to meet, they also are called converging lenses.

The image formed by a convex lens is similar to the image formed by a concave mirror. For both, the type of image depends on how far the object is from the mirror or lens. Look at **Figure 18B.** If the object is farther than two focal lengths from the lens, the image seen through the lens is inverted and smaller than the object.

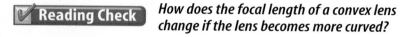

Reading Check *How does the focal length of a convex lens change if the lens becomes more curved?*

If the object is closer to the lens than one focal length, then the image formed is right-side up and larger than the object, as shown in **Figure 18C.** A magnifying glass forms an image in this way. As long as the magnifying glass is less than one focal length from the object, you can make the image appear larger by moving the magnifying glass away from the object.

Concave Lenses

A lens that is thicker at the edges than in the middle is a **concave lens**. A concave lens also is called a diverging lens. **Figure 19** shows how light rays traveling parallel to the optical axis are bent after passing through a concave lens.

A concave lens causes light rays to diverge, so light rays are not brought to a focus. The type of image that is formed by a concave lens is similar to one that is formed by a convex mirror. The image is upright and smaller than the object.

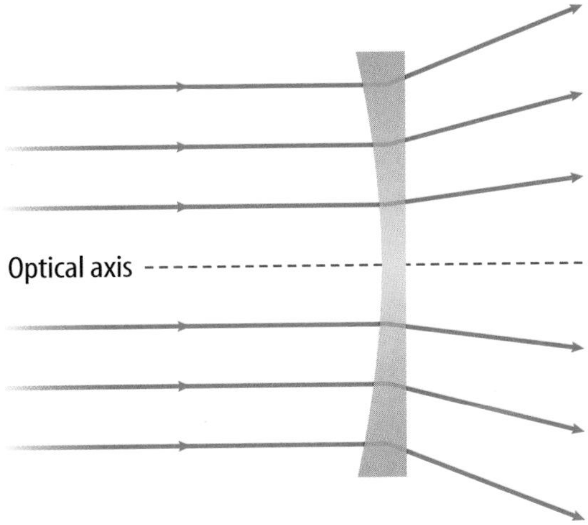

Total Internal Reflection

When you look at a glass window, you sometimes can see your reflection. You see a reflection because some of the light waves reflected from you are reflected back to your eyes when they strike the window. This is an example of a partial reflection—only some of the light waves striking the window are reflected. However, sometimes all the light waves that strike the boundary between two transparent materials can be reflected. This process is called total internal reflection.

The Critical Angle To see how total internal reflection occurs, look at **Figure 20.** Light travels faster in air than in water, and the refracted beam is bent away from the normal. As the angle between the incident beam and the normal increases, the refracted beam bends closer to the air-water boundary. At the same time, more of the light energy striking the boundary is reflected and less light energy passes into the air.

If a light beam in water strikes the boundary so that the angle with the normal is greater than an angle called the critical angle, total internal reflection occurs. Then all the light waves are reflected at the air-water boundary, just as if a mirror were there. The size of the critical angle depends on the two materials involved. For light passing from water to air, the critical angle is about 48 degrees.

Figure 19 A concave lens causes light rays traveling parallel to the optical axis to diverge.

Figure 20 When a light beam passes from one medium to another, some of its energy is reflected (red) and some is refracted (blue).

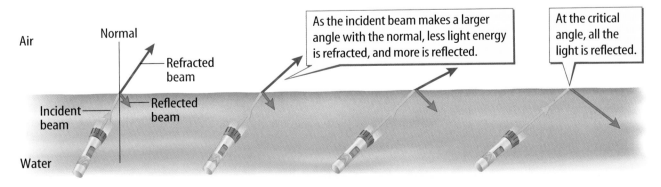

As the incident beam makes a larger angle with the normal, less light energy is refracted, and more is reflected.

At the critical angle, all the light is reflected.

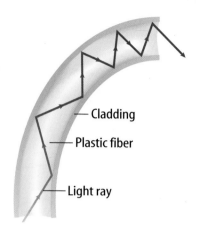

Figure 21 An optical fiber is made of materials that cause total internal reflection to occur. A light beam can travel for many kilometers through an optical fiber and lose almost no energy.

Cladding

Plastic fiber

Light ray

Optical Fibers Optical fibers are thin, flexible, transparent fibers. An optical fiber is like a light pipe. Even if the fiber is bent, light that enters one end of the fiber comes out the other end.

Total internal reflection makes light transmission in optical fibers possible. A thin fiber of glass or plastic is covered with another material called cladding in which light travels faster. When light strikes the boundary between the fiber and the cladding, total internal reflection can occur. In this way, the beam bounces along inside the fiber as shown in **Figure 21.**

Optical fibers are used most commonly in the communications industry. For example, television programs, computer information, and phone conversations can be coded into light signals. These signals then can be sent from one place to another using optical fibers. Because of total internal reflection, signals can't leak from one fiber to another and interfere with others. As a result, the signal is transmitted clearly. One optical fiber the thickness of a human hair can carry thousands of phone conversations.

section 3 review

Summary

The Refraction of Light

- Light travels at different speeds in different materials.
- Refraction occurs when light changes speed as it travels from one material into another.

Convex and Concave Lenses

- A lens is a transparent object with at least one curved side that causes light to bend.
- A convex lens is thicker in the center than at the edges and causes light waves to converge.
- A concave lens is thinner in the center than at the edges and causes light waves to diverge.

Total Internal Reflection

- Total internal reflection occurs at the boundary between two transparent materials when light is completely reflected.
- Optical fibers use total internal reflection to transmit information over long distances with light waves.

Self Check

1. **Compare** the image formed by a concave lens and the image formed by a convex mirror.
2. **Explain** whether you would use a convex lens or a concave lens to magnify an object.
3. **Describe** the image formed by convex lens if an object is less than one focal length from the lens.
4. **Describe** how light rays traveling parallel to the optical axis are bent after they pass through a convex lens.
5. **Infer** If the speed of light were the same in all materials, would a lens cause light rays to bend?
6. **Think Critically** A light wave is bent more when it travels from air to glass than when it travels from air to water. Is the speed of light greater in water or in glass? Explain.

Applying Math

7. **Calculate Time** If light travels at 300,000 km/s and Earth is 150 million km from the Sun, how long does it take light to travel form the Sun to Earth?

Using Mirrors and Lenses

Microscopes

For almost 500 years, lenses have been used to observe objects that are too small to be seen with the unaided eye. The first microscopes were simple and magnified less than 100 times. Today, a compound microscope like the one in **Figure 22** uses a combination of lenses to magnify objects by as much as 2,500 times.

Figure 22 also shows how a microscope forms an image. An object, such as an insect or a drop of water from a pond, is placed close to a convex lens called the objective lens. This lens produces an enlarged image inside the microscope tube. The light rays from that image then pass through a second convex lens called the eyepiece lens. This lens further magnifies the image formed by the objective lens. By using two lenses, a much larger image is formed than a single lens can produce.

as you read

What You'll Learn

- **Explain** how microscopes magnify objects.
- **Explain** how telescopes make distant objects visible.
- **Describe** how a camera works.

Why It's Important

Microscopes and telescopes are used to view parts of the universe that can't be seen with the unaided eye.

Review Vocabulary

retina: region on the inner surface of the back of the eye that contains light-sensitive cells

New Vocabulary

- refracting telescope
- reflecting telescope

Figure 22 A compound microscope uses lenses to magnify objects.

A compound microscope often has more than one objective lens—each providing a different magnification. A light underneath the objective lens makes the image bright enough to see clearly.

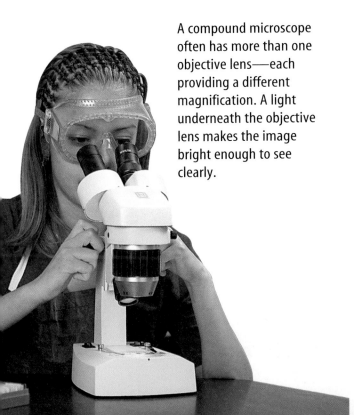

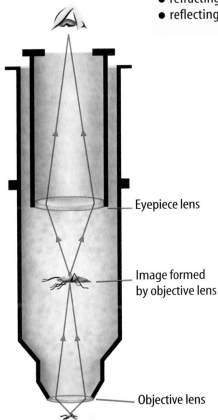

Eyepiece lens

Image formed by objective lens

Objective lens

Object

The objective lens in a compound microscope forms an enlarged image, which is then magnified by the eyepiece lens.

Telescopes

Just as microscopes are used to magnify very small objects, telescopes are used to examine objects that are very far away. The first telescopes were made at about the same time as the first microscopes. Much of what is known about the Moon, the solar system, and the distant universe has come from images and other information gathered by telescopes.

Refracting Telescopes The simplest **refracting telescopes** use two convex lenses to form an image of a distant object. Just as in a compound microscope, light passes through an objective lens that forms an image. That image is then magnified by an eyepiece, as shown in **Figure 23.**

An important difference between a telescope and a microscope is the size of the objective lens. The main purpose of a telescope is not to magnify an image. A telescope's main purpose is to gather as much light as possible from distant objects. The larger an objective lens is, the more light can enter it. This makes images of faraway objects look brighter and more detailed when they are magnified by the eyepiece. With a large enough objective lens, it's possible to see stars and galaxies that are many trillions of kilometers away. **Figure 23** also shows the largest refracting telescope ever made.

Reading Check *How does a telescope's objective lens enable distant objects to be seen?*

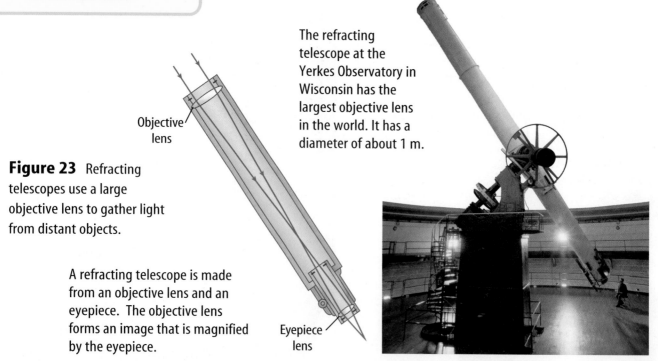

Objective lens

Eyepiece lens

Figure 23 Refracting telescopes use a large objective lens to gather light from distant objects.

A refracting telescope is made from an objective lens and an eyepiece. The objective lens forms an image that is magnified by the eyepiece.

The refracting telescope at the Yerkes Observatory in Wisconsin has the largest objective lens in the world. It has a diameter of about 1 m.

Figure 24 Reflecting telescopes gather light by using a concave mirror.

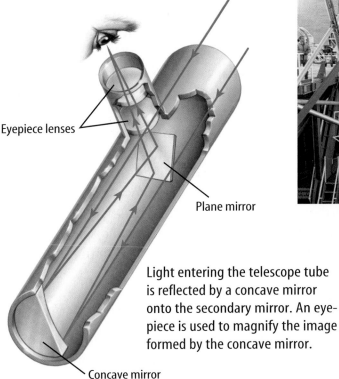

Eyepiece lenses

Plane mirror

Light entering the telescope tube is reflected by a concave mirror onto the secondary mirror. An eyepiece is used to magnify the image formed by the concave mirror.

Concave mirror

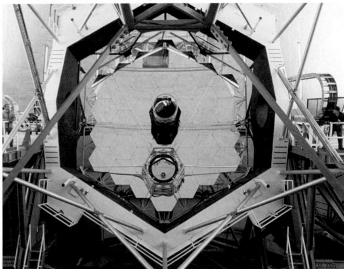

The Keck telescope in Mauna Kea, Hawaii, is the largest reflecting telescope in the world.

Reflecting Telescopes Refracting telescopes have size limitations. One problem is that the objective lens can be supported only around its edges. If the lens is extremely large, it cannot be supported enough to keep the glass from sagging slightly under its own weight. This causes the image that the lens forms to become distorted.

Reflecting telescopes can be made much larger than refracting telescopes. **Reflecting telescopes** have a concave mirror instead of a concave objective lens to gather the light from distant objects. As shown in **Figure 24**, the large concave mirror focuses light onto a secondary mirror that directs it to the eyepiece, which magnifies the image.

Because only the one reflecting surface on the mirror needs to be made carefully and kept clean, telescope mirrors are less expensive to make and maintain than lenses of a similar size. Also, mirrors can be supported not only at their edges but also on their backsides. They can be made much larger without sagging under their own weight. The Keck telescope in Hawaii, shown in **Figure 24,** is the largest reflecting telescope in the world. Its large concave mirror is 10 m in diameter, and is made of 36 six-sided segments. Each segment is 1.8 m in size and the segments are pieced together to form the mirror.

The First Telescopes
A Dutch eyeglass maker, Hans Lippershey, constructed a refracting telescope in 1608 that had a magnification of 3. In 1609 Galileo built a refracting telescope with a magnification of 20. By 1668, the first reflecting telescope was built by Isaac Newton that had a metal concave mirror with a diameter of about 5 cm. More than a century later, William Herschel built the first large reflecting telescopes with mirrors as large as 50 cm. Research the history of the telescope and make a timeline showing important events.

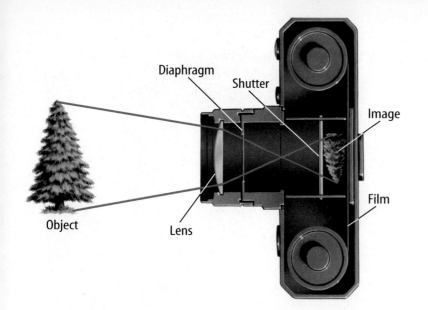

Diaphragm
Shutter
Image
Film
Object
Lens

Figure 25 A camera uses a convex lens to form an image on a piece of light-sensitive film. The image formed by a camera lens is smaller than the object and is inverted.

Science nline

Topic: Lasers
Visit booko.msscience.com for Web links to information about uses for lasers.

Activity Make a table listing different types of lasers and how they are used.

Cameras

You probably see photographs taken by cameras almost every day. A typical camera uses a convex lens to form an image on a section of film, just as your eye's lens focuses an image on your retina. The convex lens has a short focal length, so it forms an image that is smaller than the object and inverted on the film. Look at the camera shown in **Figure 25.** When the shutter is open, the convex lens focuses an image on a piece of film that is sensitive to light. Light-sensitive film contains chemicals that undergo chemical reactions when light hits it. The brighter parts of the image affect the film more than the darker parts do.

✓ Reading Check *What type of lens does a camera use?*

If too much light strikes the film, the image formed on the film is overexposed and looks washed out. On the other hand, if too little light reaches the film, the photograph might be too dark. To control how much light reaches the film, many cameras have a device called a diaphragm. The diaphragm is opened to let more light onto the film and closed to reduce the amount of light that strikes the film.

Lasers

Perhaps you've seen the narrow, intense beams of laser light used in a laser light show. Intense laser beams are also used for different kinds of surgery. Why can laser beams be so intense? One reason is that a laser beam doesn't spread out as much as ordinary light as it travels.

Spreading Light Beams Suppose you shine a flashlight on a wall in a darkened room. The size of the spot of light on the wall depends on the distance between the flashlight and the wall. As the flashlight moves farther from the wall, the spot of light gets larger. This is because the beam of light produced by the flashlight spreads out as it travels. As a result, the energy carried by the light beam is spread over an increasingly larger area as the distance from the flashlight gets larger. As the energy is spread over a larger area, the energy becomes less concentrated and the intensity of the beam decreases.

Using Laser Light Laser light is different from the light produced by the flashlight in several ways, as shown in **Figure 26.** One difference is that in a beam of laser light, the crests and troughs of the light waves overlap, so the waves are in phase.

Because a laser beam doesn't spread out as much as ordinary light, a large amount of energy can be applied to a very small area. This property enables lasers to be used for cutting and welding materials and as a replacement for scalpels in surgery. Less intense laser light is used for such applications as reading and writing to CDs or in grocery store bar-code readers. Surveyors and builders use lasers to measure distances, angles, and heights. Laser beams also are used to transmit information through space or through optical fibers.

Figure 26 Laser light is different from the light produced by a lightbulb.

The light from a bulb contains waves with many different wavelengths that are out of phase and traveling in different directions.

The light from a laser contains waves with only one wavelength that are in phase and traveling in the same direction.

section 4 review

Summary

Microscopes, Telescopes, and Cameras

- A compound microscope uses an objective lens and an eyepiece lens to form an enlarged image of an object.
- A refracting telescope contains a large objective lens to gather light and a smaller eyepiece lens to magnify the image.
- A reflecting telescope uses a large concave mirror to gather light and an eyepiece lens to magnify the image.
- The image formed by a telescope becomes brighter and more detailed as the size of the objective lens or concave mirror increases.
- A camera uses a convex lens to form an image on light-sensitive film.

Laser Light

- Light from a laser contains light waves that are in phase, have only one wavelength, and travel in the same direction.
- Because laser light does not spread out much as it travels the energy it carries can be applied over a very small area.

Self Check

1. **Explain** why the concave mirror of a reflecting telescope can be made much larger than the objective lens of a refracting telescope.
2. **Describe** how a beam of laser light is different than the beam of light produced by a flashlight.
3. **Explain** why the objective lens of a refracting telescope is much larger than the objective lens of a compound microscope.
4. **Infer** how the image produced by a compound microscope would be different if the eyepiece lens were removed from the microscope.
5. **Think Critically** Explain why the intensity of the light in a flashlight beam decreases as the flashlight moves farther away.

Applying Math

6. **Calculate Image Size** The size of an image is related to the magnification of an optical instrument by the following formula:

    Image size = magnification × object size

   A blood cell has a diameter of 0.001 cm. How large is the image formed by a microscope with a magnification of 1,000?

Image Formation by a Convex Lens

Goals

- **Measure** the image distance as the object distance changes.
- **Observe** the type of image formed as the object distance changes.

Possible Materials

convex lens
modeling clay
meterstick
flashlight
masking tape
20-cm square piece of cardboard with a white surface

Safety Precautions

▶ Real-World Question

The type of image formed by a convex lens, also called a converging lens, is related to the distance of the object from the lens. This distance is called the object distance. The location of the image also is related to the distance of the object from the lens. The distance from the lens to the image is called the image distance. How are the image distance and object distance related for a convex lens?

▶ Procedure

1. **Design** a data table to record your data. Make three columns in your table —one column for the object distance, another for the image distance, and the third for the type of image.

Convex Lens Data		
Object Distance (m)	Image Distance (m)	Image Type
Do not write in this book.		

2. Use the modeling clay to make the lens stand upright on the lab table.
3. Form the letter *F* on the glass surface of the flashlight with masking tape.
4. Turn on the flashlight and place it 1 m from the lens. Position the flashlight so the flashlight beam is shining through the lens.
5. **Record** the distance from the flashlight to the lens in the object distance column in your data table.
6. Hold the cardboard vertically upright on the other side of the lens, and move it back and forth until a sharp image of the letter *F* is obtained.

7. **Measure** the distance of the card from the lens using the meterstick, and record this distance in the Image Distance column in your data table.

8. **Record** in the third column of your data table whether the image is upright or inverted, and smaller or larger.

9. Repeat steps 4 through 8 for object distances of 0.50 m and 0.25 m and record your data in your data table.

◉ Analyze Your Data

1. **Describe** any observed relationship between the object distance, and the image type.

2. **Identify** the variables involved in determining the image type for a convex lens.

◉ Conclude and Apply

1. **Explain** how the image distance changed as the object distance decreased.

2. **Identify** how the image changed as the object distance decreased.

3. **Predict** what would happen to the size of the image if the flashlight were much farther away than 1 m.

Communicating Your Data

Demonstrate this lab to a third-grade class and explain how it works. **For more help, refer to the** Science Skill Handbook.

Oops! Accidents in SCIENCE

Eyeglasses

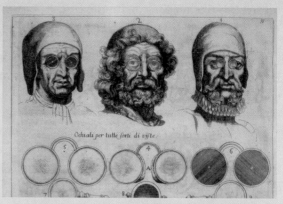

This Italian engraving from the 1600s shows some different types of glasses.

"**I**t is not yet twenty years since the art of making spectacles, one of the most useful arts on Earth, was discovered. I, myself, have seen and conversed with the man who made them first."

This quote from an Italian monk dates back to 1306 and is one of the first historical records to refer to eyeglasses. Unfortunately, the monk, Giordano, never actually named the man he met. Thus, the inventor of eyeglasses remains unknown.

The mystery exists, in part, because different cultures in different places used some type of magnifying tool to improve their vision. For example, a rock-crystal lens, made by early Assyrians who lived 3,500 years ago in what is now Iraq, may have been used to improve vision. About 2,000 years ago, the Roman writer Seneca looked through a glass globe of water to make the letters appear bigger in the books he read. By the tenth century, glasses had been invented in China, but they were used to keep away bad luck, not to improve vision.

In the mid 1400s in Europe, eyeglasses began to appear in paintings of scholars, clergy, and the upper classes—eyeglasses were so expensive that only the rich could afford them. In the early 1700s, for example, glasses cost roughly $200, which is comparable to thousands of dollars today. By the mid-1800s, improvements in manufacturing techniques made eyeglasses much less expensive to make, and thus this important invention became widely available to people of all walks of life.

How Eyeglasses Work

Eyeglasses are used to correct farsightedness and nearsightedness, as well as other vision problems. The eye focuses light rays to form an image on a region called the retina on the back of the eye. Farsighted people have difficulty seeing things close up because light rays from nearby objects do not converge enough to form an image on the retina. This problem can be corrected by using convex lenses that cause light rays to converge before they enter the eye. Nearsighted people have problems seeing distant objects because light rays from far-away objects are focused in front of the retina. Concave lenses that cause light rays to diverge are used to correct this vision problem.

Research In many parts of the world, people have no vision care, and eye diseases and poor vision go untreated. Research the work of groups that bring eye care to people.

Reviewing Main Ideas

Section 1 Properties of Light

1. Light waves can be absorbed, reflected, or transmitted when they strike an object.

2. The color of an object depends on the wavelengths of light reflected by the object.

Section 2 Reflection and Mirrors

1. Light reflected from the surface of an object obeys the law of reflection—the angle of incidence equals the angle of reflection.

2. Concave mirrors cause light waves to converge, or meet. Convex mirrors cause light waves to diverge, or spread apart.

Section 3 Refraction and Lenses

1. Light waves bend, or refract, when they change speed in traveling from one medium to another.

2. A convex lens causes light waves to converge, and a concave lens causes light waves to diverge.

Section 4 Using Mirrors and Lenses

1. A compound microscope uses a convex objective lens to form an enlarged image that is further enlarged by an eyepiece.

2. A refracting telescope uses a large objective lens and an eyepiece lens to form an image of a distant object.

3. A reflecting telescope uses a large concave mirror that gathers light and an eyepiece lens to form an image of a distant object.

4. Cameras use a convex lens to form an image on light-sensitive film.

Visualizing Main Ideas

Copy and complete the following concept map.

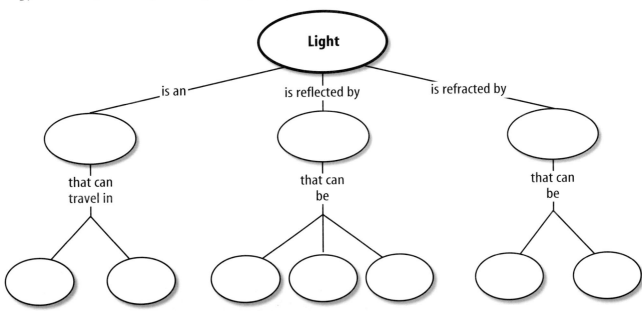

chapter 4 Review

Using Vocabulary

concave lens p. 111	lens p. 109
convex lens p. 110	light ray p. 96
focal length p. 104	medium p. 97
focal point p. 104	reflecting telescope p. 115
law of reflection p. 101	refracting telescope p. 114

Complete each statement using a word or words from the vocabulary list above.

1. A _____ is the material in which a light wave travels.

2. A narrow beam of light that travels in a straight line is a _____.

3. The _____ is the distance from a lens or a mirror to the focal point.

4. Light rays traveling parallel to the optical axis of a convex lens are bent so they pass through the _____.

5. A transparent object with at least one curved surface that causes light waves to bend is a _____.

6. A _____ is thicker in the center than it is at the edges.

7. A _____ uses a large concave mirror to gather light from distant objects.

Checking Concepts

Choose the word or phrase that best answers the question.

8. Light waves travel the fastest through which of the following?
   A) air
   B) diamond
   C) water
   D) a vacuum

9. Which of the following determines the color of light?
   A) a prism
   B) its refraction
   C) its wavelength
   D) its incidence

10. If an object reflects red and green light, what color does the object appear to be?
   A) yellow
   B) red
   C) green
   D) purple

11. If an object absorbs all the light that hits it, what color is it?
   A) white
   B) blue
   C) black
   D) green

12. What type of image is formed by a plane mirror?
   A) upright
   B) inverted
   C) magnified
   D) all of these

13. How is the angle of incidence related to the angle of reflection?
   A) It's greater.
   B) It's smaller.
   C) It's the same.
   D) It's not focused.

14. Which of the following can be used to magnify objects?
   A) a concave lens
   B) a convex lens
   C) a convex mirror
   D) all of these

15. Which of the following describes the light waves that make up laser light?
   A) same wavelength
   B) same direction
   C) in phase
   D) all of these

16. What is an object that reflects some light and transmits some light called?
   A) colored
   B) diffuse
   C) opaque
   D) translucent

17. What is the main purpose of the objective lens or concave mirror in a telescope?
   A) invert images
   B) reduce images
   C) gather light
   D) magnify images

18. Which of the following types of mirror can form an image larger than the object?
   A) convex
   B) concave
   C) plane
   D) all of these

Science Online   booko.msscience.com/vocabulary_puzzlemaker

Thinking Critically

19. Diagram Suppose you can see a person's eyes in a mirror. Draw a diagram to determine whether or not that person can see you.

20. Determine A singer is wearing a blue outfit. What color spotlights could be used to make the outfit appear to be black?

21. Form a hypothesis to explain why sometimes you can see two images of yourself reflected from a window at night.

22. Explain why a rough surface, such as a road, becomes shiny in appearance and a better reflector when it is wet.

23. Infer An optical fiber is made of a material that forms the fiber and a different material that forms the outer covering. For total internal reflection to occur, how does the speed of light in the fiber compare with the speed of light in the outer covering?

Use the table below to answer question 24.

Magnification by a Convex Lens	
Object Distance (cm)	Magnification
25	4.00
30	2.00
40	1.00
60	0.50
100	0.25

24. Use a Table In the table above, the object distance is the distance of the object from the lens. The magnification is the image size divided by the object size. If the focal length of the lens is 20 cm, how does the size of the image change as the object gets farther from the focal point?

25. Calculate What is the ratio of the distance at which the magnification equals 1.00 to the focal length of the lens?

Performance Activities

26. Oral Presentation Investigate the types of mirrors used in fun houses. Explain how these mirrors are formed, and why they produce distorted images. Demonstrate your findings to your class.

27. Reverse Writing Images are reversed left to right in a plane mirror. Write a note to a friend that can be read only in a plane mirror.

28. Design an experiment to determine the focal length of a convex lens. Write a report describing your experiment, including a diagram.

Applying Math

Use the graph below to answer questions 29 and 30.

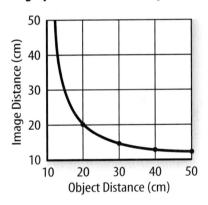

29. Image Position The graph shows how the distance of an image from a convex lens is related to the distance of the object from the lens. How does the position of the image change as the object gets closer to the lens?

30. Magnification The magnification of the image equals the image distance divided by the object distance. At what object distance does the magnification equal 2?

Part 1 Multiple Choice

Record your answers on the answer sheet provided by your teacher or on a sheet of paper.

1. Which of the following describes an object that allows no light to pass through it?
   A. transparent   C. opaque
   B. translucent   D. diffuse

2. Which statement is always true about the image formed by a concave lens?
   A. It is upside down and larger than the object.
   B. It is upside down and smaller than the object.
   C. It is upright and larger than the object.
   D. It is upright and smaller than the object.

Use the figure below to answer questions 3 and 4.

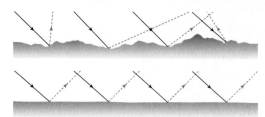

3. Which of the following describes the process occurring in the upper panel of the figure?
   A. refraction
   B. diffuse reflection
   C. regular reflection
   D. total internal reflection

4. The surface in the lower panel of the figure would be like which of the following?
   A. a mirror   C. a sheet of paper
   B. waxed paper   D. a painted wall

5. Why does a leaf look green?
   A. It reflects green light.
   B. It absorbs green light.
   C. It reflects all colors of light.
   D. It reflects all colors except green.

6. What does a refracting telescope use to form an image of a distant object?
   A. two convex lenses
   B. a concave mirror and a plane mirror
   C. two concave lenses
   D. two concave mirrors

7. Through which of the following does light travel the slowest?
   A. air   C. water
   B. diamond   D. vacuum

8. What is the bending of a light wave due to a change in speed?
   A. reflection   C. refraction
   B. diffraction   D. transmission

Use the figure below to answer questions 9 and 10.

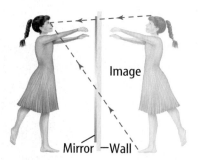

Image

Mirror ─Wall

9. If the girl is standing 1 m from the mirror, where will her image seem to be located?
   A. 2 m behind the mirror
   B. 1 m behind the mirror
   C. 2 m in front of the mirror
   D. 1 m in front of the mirror

10. Which of the following describes the image of the girl formed by the plane mirror?
    A. It will be upside down.
    B. It will be in front of the mirror.
    C. It will be larger than the girl.
    D. It will be reversed left to right.

Part 2 | Short Response/Grid In

Record your answers on the answer sheet provided by your teacher or on a sheet of paper.

11. Light travels slower in diamond than in air. Explain whether total internal reflection could occur for a light wave traveling in the diamond toward the diamond's surface.

Use the figure below to answer question 12.

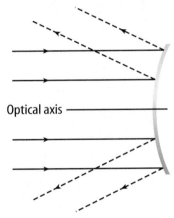

Optical axis

12. Identify the type of mirror shown in the figure and describe the image this mirror forms.

13. Under white light the paper of this page looks white and the print looks black. What color would the paper and the print appear to be under red light?

14. A light ray strikes a plane mirror such that the angle of incidence is 30°. What is the angle between the light ray and the surface of the mirror?

15. Contrast the light beam from a flashlight and a laser light beam.

16. An actor on stage is wearing a magenta outfit. Explain what color the outfit would appear in red light, in blue light, and in green light.

17. To use a convex lens as a magnifying lens, where must the object be located?

Part 3 | Open Ended

Record your answers on a sheet of paper.

18. Explain why you can see the reflection of trees in the water of a lake on a calm day, but not a very windy day.

19. Describe how total internal reflection enables optical fibers to transmit light over long distances.

20. What happens when a source of light is placed at the focal point of a concave mirror? Give an example.

Use the illustration below to answer question 21.

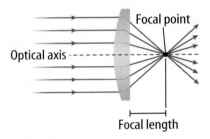

Focal point

Optical axis

Focal length

21. Describe how the position of the focal point changes as the lens becomes flatter and less curved.

22. Compare the images formed by a concave mirror when an object is between the focal point and the mirror and when an object is beyond the focal point.

23. Explain why increasing the size of the concave mirror in a reflecting telescope improves the images formed.

Test-Taking Tip

Organize Your Main Points For essay questions, spend a few minutes listing and organizing the main points that you plan to discuss. Make sure to do all of this work on your scratch paper, not your answer sheet.

Question 19 Organize your discussion points by first listing what you know about optical fibers and total internal reflection.

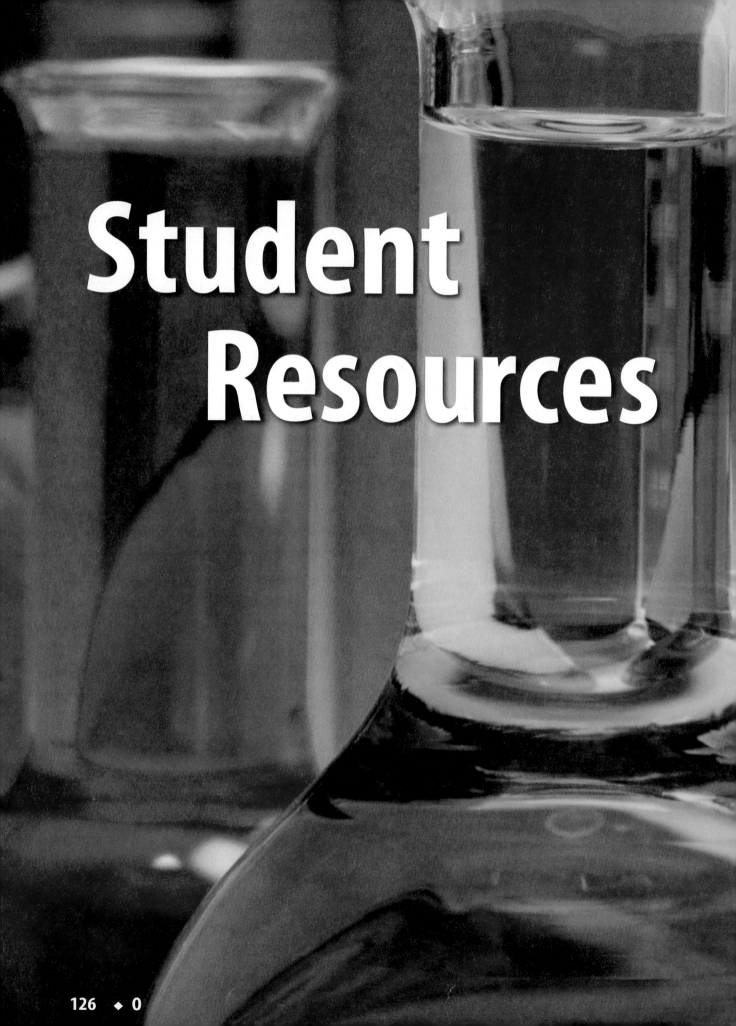

Student Resources

CONTENTS

Scientific Methods

Scientists use an orderly approach called the scientific method to solve problems. This includes organizing and recording data so others can understand them. Scientists use many variations in this method when they solve problems.

Identify a Question

The first step in a scientific investigation or experiment is to identify a question to be answered or a problem to be solved. For example, you might ask which gasoline is the most efficient.

Gather and Organize Information

After you have identified your question, begin gathering and organizing information. There are many ways to gather information, such as researching in a library, interviewing those knowledgeable about the subject, testing and working in the laboratory and field. Fieldwork is investigations and observations done outside of a laboratory.

Researching Information Before moving in a new direction, it is important to gather the information that already is known about the subject. Start by asking yourself questions to determine exactly what you need to know. Then you will look for the information in various reference sources, like the student is doing in **Figure 1.** Some sources may include textbooks, encyclopedias, government documents, professional journals, science magazines, and the Internet. Always list the sources of your information.

Figure 1 The Internet can be a valuable research tool.

Evaluate Sources of Information Not all sources of information are reliable. You should evaluate all of your sources of information, and use only those you know to be dependable. For example, if you are researching ways to make homes more energy efficient, a site written by the U.S. Department of Energy would be more reliable than a site written by a company that is trying to sell a new type of weatherproofing material. Also, remember that research always is changing. Consult the most current resources available to you. For example, a 1985 resource about saving energy would not reflect the most recent findings.

Sometimes scientists use data that they did not collect themselves, or conclusions drawn by other researchers. This data must be evaluated carefully. Ask questions about how the data were obtained, if the investigation was carried out properly, and if it has been duplicated exactly with the same results. Would you reach the same conclusion from the data? Only when you have confidence in the data can you believe it is true and feel comfortable using it.

Interpret Scientific Illustrations As you research a topic in science, you will see drawings, diagrams, and photographs to help you understand what you read. Some illustrations are included to help you understand an idea that you can't see easily by yourself, like the tiny particles in an atom in **Figure 2.** A drawing helps many people to remember details more easily and provides examples that clarify difficult concepts or give additional information about the topic you are studying. Most illustrations have labels or a caption to identify or to provide more information.

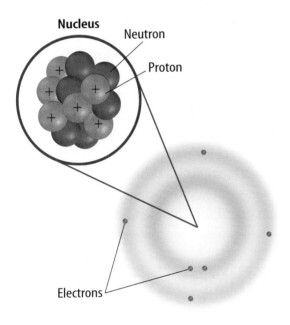

Figure 2 This drawing shows an atom of carbon with its six protons, six neutrons, and six electrons.

Concept Maps One way to organize data is to draw a diagram that shows relationships among ideas (or concepts). A concept map can help make the meanings of ideas and terms more clear, and help you understand and remember what you are studying. Concept maps are useful for breaking large concepts down into smaller parts, making learning easier.

Network Tree A type of concept map that not only shows a relationship, but how the concepts are related is a network tree, shown in **Figure 3.** In a network tree, the words are written in the ovals, while the description of the type of relationship is written across the connecting lines.

When constructing a network tree, write down the topic and all major topics on separate pieces of paper or notecards. Then arrange them in order from general to specific. Branch the related concepts from the major concept and describe the relationship on the connecting line. Continue to more specific concepts until finished.

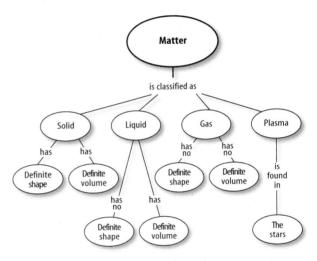

Figure 3 A network tree shows how concepts or objects are related.

Events Chain Another type of concept map is an events chain. Sometimes called a flow chart, it models the order or sequence of items. An events chain can be used to describe a sequence of events, the steps in a procedure, or the stages of a process.

When making an events chain, first find the one event that starts the chain. This event is called the initiating event. Then, find the next event and continue until the outcome is reached, as shown in **Figure 4.**

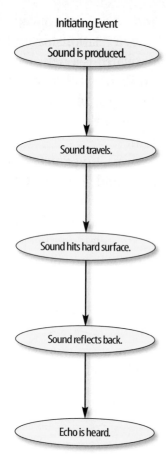

Initiating Event

Sound is produced.

Sound travels.

Sound hits hard surface.

Sound reflects back.

Echo is heard.

Figure 4 Events-chain concept maps show the order of steps in a process or event. This concept map shows how a sound makes an echo.

Cycle Map A specific type of events chain is a cycle map. It is used when the series of events do not produce a final outcome, but instead relate back to the beginning event, such as in **Figure 5.** Therefore, the cycle repeats itself.

To make a cycle map, first decide what event is the beginning event. This is also called the initiating event. Then list the next events in the order that they occur, with the last event relating back to the initiating event. Words can be written between the events that describe what happens from one event to the next. The number of events in a cycle map can vary, but usually contain three or more events.

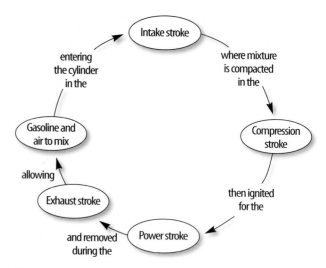

Figure 5 A cycle map shows events that occur in a cycle.

Spider Map A type of concept map that you can use for brainstorming is the spider map. When you have a central idea, you might find that you have a jumble of ideas that relate to it but are not necessarily clearly related to each other. The spider map on sound in **Figure 6** shows that if you write these ideas outside the main concept, then you can begin to separate and group unrelated terms so they become more useful.

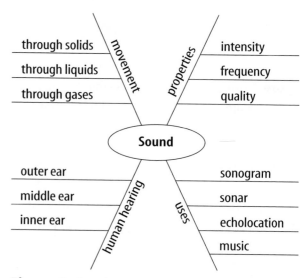

Figure 6 A spider map allows you to list ideas that relate to a central topic but not necessarily to one another.

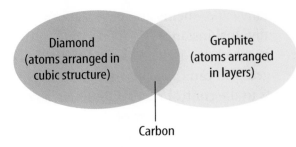

Figure 7 This Venn diagram compares and contrasts two substances made from carbon.

Venn Diagram To illustrate how two subjects compare and contrast you can use a Venn diagram. You can see the characteristics that the subjects have in common and those that they do not, shown in **Figure 7.**

To create a Venn diagram, draw two overlapping ovals that that are big enough to write in. List the characteristics unique to one subject in one oval, and the characteristics of the other subject in the other oval. The characteristics in common are listed in the overlapping section.

Make and Use Tables One way to organize information so it is easier to understand is to use a table. Tables can contain numbers, words, or both.

To make a table, list the items to be compared in the first column and the characteristics to be compared in the first row. The title should clearly indicate the content of the table, and the column or row heads should be clear. Notice that in **Table 1** the units are included.

Table 1 Recyclables Collected During Week			
Day of Week	**Paper (kg)**	**Aluminum (kg)**	**Glass (kg)**
Monday	5.0	4.0	12.0
Wednesday	4.0	1.0	10.0
Friday	2.5	2.0	10.0

Make a Model One way to help you better understand the parts of a structure, the way a process works, or to show things too large or small for viewing is to make a model. For example, an atomic model made of a plastic-ball nucleus and pipe-cleaner electron shells can help you visualize how the parts of an atom relate to each other. Other types of models can by devised on a computer or represented by equations.

Form a Hypothesis

A possible explanation based on previous knowledge and observations is called a hypothesis. After researching gasoline types and recalling previous experiences in your family's car you form a hypothesis—our car runs more efficiently because we use premium gasoline. To be valid, a hypothesis has to be something you can test by using an investigation.

Predict When you apply a hypothesis to a specific situation, you predict something about that situation. A prediction makes a statement in advance, based on prior observation, experience, or scientific reasoning. People use predictions to make everyday decisions. Scientists test predictions by performing investigations. Based on previous observations and experiences, you might form a prediction that cars are more efficient with premium gasoline. The prediction can be tested in an investigation.

Design an Experiment A scientist needs to make many decisions before beginning an investigation. Some of these include: how to carry out the investigation, what steps to follow, how to record the data, and how the investigation will answer the question. It also is important to address any safety concerns.

Test the Hypothesis

Now that you have formed your hypothesis, you need to test it. Using an investigation, you will make observations and collect data, or information. This data might either support or not support your hypothesis. Scientists collect and organize data as numbers and descriptions.

Follow a Procedure In order to know what materials to use, as well as how and in what order to use them, you must follow a procedure. **Figure 8** shows a procedure you might follow to test your hypothesis.

Procedure

1. Use regular gasoline for two weeks.
2. Record the number of kilometers between fill-ups and the amount of gasoline used.
3. Switch to premium gasoline for two weeks.
4. Record the number of kilometers between fill-ups and the amount of gasoline used.

Figure 8 A procedure tells you what to do step by step.

Identify and Manipulate Variables and Controls In any experiment, it is important to keep everything the same except for the item you are testing. The one factor you change is called the independent variable. The change that results is the dependent variable. Make sure you have only one independent variable, to assure yourself of the cause of the changes you observe in the dependent variable. For example, in your gasoline experiment the type of fuel is the independent variable. The dependent variable is the efficiency.

Many experiments also have a control—an individual instance or experimental subject for which the independent variable is not changed. You can then compare the test results to the control results. To design a control you can have two cars of the same type. The control car uses regular gasoline for four weeks. After you are done with the test, you can compare the experimental results to the control results.

Collect Data

Whether you are carrying out an investigation or a short observational experiment, you will collect data, as shown in **Figure 9.** Scientists collect data as numbers and descriptions and organize it in specific ways.

Observe Scientists observe items and events, then record what they see. When they use only words to describe an observation, it is called qualitative data. Scientists' observations also can describe how much there is of something. These observations use numbers, as well as words, in the description and are called quantitative data. For example, if a sample of the element gold is described as being "shiny and very dense" the data are qualitative. Quantitative data on this sample of gold might include "a mass of 30 g and a density of 19.3 g/cm^3."

Figure 9 Collecting data is one way to gather information directly.

Figure 10 Record data neatly and clearly so it is easy to understand.

When you make observations you should examine the entire object or situation first, and then look carefully for details. It is important to record observations accurately and completely. Always record your notes immediately as you make them, so you do not miss details or make a mistake when recording results from memory. Never put unidentified observations on scraps of paper. Instead they should be recorded in a notebook, like the one in **Figure 10.** Write your data neatly so you can easily read it later. At each point in the experiment, record your observations and label them. That way, you will not have to determine what the figures mean when you look at your notes later. Set up any tables that you will need to use ahead of time, so you can record any observations right away. Remember to avoid bias when collecting data by not including personal thoughts when you record observations. Record only what you observe.

Estimate Scientific work also involves estimating. To estimate is to make a judgment about the size or the number of something without measuring or counting. This is important when the number or size of an object or population is too large or too difficult to accurately count or measure.

Sample Scientists may use a sample or a portion of the total number as a type of estimation. To sample is to take a small, representative portion of the objects or organisms of a population for research. By making careful observations or manipulating variables within that portion of the group, information is discovered and conclusions are drawn that might apply to the whole population. A poorly chosen sample can be unrepresentative of the whole. If you were trying to determine the rainfall in an area, it would not be best to take a rainfall sample from under a tree.

Measure You use measurements everyday. Scientists also take measurements when collecting data. When taking measurements, it is important to know how to use measuring tools properly. Accuracy also is important.

Length To measure length, the distance between two points, scientists use meters. Smaller measurements might be measured in centimeters or millimeters.

Length is measured using a metric ruler or meter stick. When using a metric ruler, line up the 0-cm mark with the end of the object being measured and read the number of the unit where the object ends. Look at the metric ruler shown in **Figure 11.** The centimeter lines are the long, numbered lines, and the shorter lines are millimeter lines. In this instance, the length would be 4.50 cm.

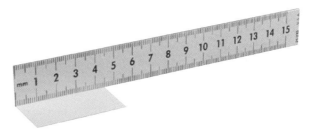

Figure 11 This metric ruler has centimeter and millimeter divisions.

Mass The SI unit for mass is the kilogram (kg). Scientists can measure mass using units formed by adding metric prefixes to the unit gram (g), such as milligram (mg). To measure mass, you might use a triple-beam balance similar to the one shown in **Figure 12.** The balance has a pan on one side and a set of beams on the other side. Each beam has a rider that slides on the beam.

When using a triple-beam balance, place an object on the pan. Slide the largest rider along its beam until the pointer drops below zero. Then move it back one notch. Repeat the process for each rider proceeding from the larger to smaller until the pointer swings an equal distance above and below the zero point. Sum the masses on each beam to find the mass of the object. Move all riders back to zero when finished.

Instead of putting materials directly on the balance, scientists often take a tare of a container. A tare is the mass of a container into which objects or substances are placed for measuring their masses. To mass objects or substances, find the mass of a clean container. Remove the container from the pan, and place the object or substances in the container. Find the mass of the container with the materials in it. Subtract the mass of the empty container from the mass of the filled container to find the mass of the materials you are using.

Figure 12 A triple-beam balance is used to determine the mass of an object.

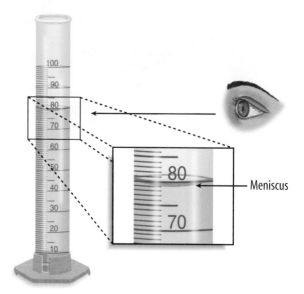

Figure 13 Graduated cylinders measure liquid volume.

Liquid Volume To measure liquids, the unit used is the liter. When a smaller unit is needed, scientists might use a milliliter. Because a milliliter takes up the volume of a cube measuring 1 cm on each side it also can be called a cubic centimeter ($cm^3 = cm \times cm \times cm$).

You can use beakers and graduated cylinders to measure liquid volume. A graduated cylinder, shown in **Figure 13,** is marked from bottom to top in milliliters. In lab, you might use a 10-mL graduated cylinder or a 100-mL graduated cylinder. When measuring liquids, notice that the liquid has a curved surface. Look at the surface at eye level, and measure the bottom of the curve. This is called the meniscus. The graduated cylinder in **Figure 13** contains 79.0 mL, or 79.0 cm^3, of a liquid.

Temperature Scientists often measure temperature using the Celsius scale. Pure water has a freezing point of 0°C and boiling point of 100°C. The unit of measurement is degrees Celsius. Two other scales often used are the Fahrenheit and Kelvin scales.

Figure 14 A thermometer measures the temperature of an object.

Scientists use a thermometer to measure temperature. Most thermometers in a laboratory are glass tubes with a bulb at the bottom end containing a liquid such as colored alcohol. The liquid rises or falls with a change in temperature. To read a glass thermometer like the thermometer in **Figure 14,** rotate it slowly until a red line appears. Read the temperature where the red line ends.

Form Operational Definitions An operational definition defines an object by how it functions, works, or behaves. For example, when you are playing hide and seek and a tree is home base, you have created an operational definition for a tree.

Objects can have more than one operational definition. For example, a ruler can be defined as a tool that measures the length of an object (how it is used). It can also be a tool with a series of marks used as a standard when measuring (how it works).

Analyze the Data

To determine the meaning of your observations and investigation results, you will need to look for patterns in the data. Then you must think critically to determine what the data mean. Scientists use several approaches when they analyze the data they have collected and recorded. Each approach is useful for identifying specific patterns.

Interpret Data The word *interpret* means "to explain the meaning of something." When analyzing data from an experiment, try to find out what the data show. Identify the control group and the test group to see whether or not changes in the independent variable have had an effect. Look for differences in the dependent variable between the control and test groups.

Classify Sorting objects or events into groups based on common features is called classifying. When classifying, first observe the objects or events to be classified. Then select one feature that is shared by some members in the group, but not by all. Place those members that share that feature in a subgroup. You can classify members into smaller and smaller subgroups based on characteristics. Remember that when you classify, you are grouping objects or events for a purpose. Keep your purpose in mind as you select the features to form groups and subgroups.

Compare and Contrast Observations can be analyzed by noting the similarities and differences between two more objects or events that you observe. When you look at objects or events to see how they are similar, you are comparing them. Contrasting is looking for differences in objects or events.

Recognize Cause and Effect A cause is a reason for an action or condition. The effect is that action or condition. When two events happen together, it is not necessarily true that one event caused the other. Scientists must design a controlled investigation to recognize the exact cause and effect.

Draw Conclusions

When scientists have analyzed the data they collected, they proceed to draw conclusions about the data. These conclusions are sometimes stated in words similar to the hypothesis that you formed earlier. They may confirm a hypothesis, or lead you to a new hypothesis.

Infer Scientists often make inferences based on their observations. An inference is an attempt to explain observations or to indicate a cause. An inference is not a fact, but a logical conclusion that needs further investigation. For example, you may infer that a fire has caused smoke. Until you investigate, however, you do not know for sure.

Apply When you draw a conclusion, you must apply those conclusions to determine whether the data supports the hypothesis. If your data do not support your hypothesis, it does not mean that the hypothesis is wrong. It means only that the result of the investigation did not support the hypothesis. Maybe the experiment needs to be redesigned, or some of the initial observations on which the hypothesis was based were incomplete or biased. Perhaps more observation or research is needed to refine your hypothesis. A successful investigation does not always come out the way you originally predicted.

Avoid Bias Sometimes a scientific investigation involves making judgments. When you make a judgment, you form an opinion. It is important to be honest and not to allow any expectations of results to bias your judgments. This is important throughout the entire investigation, from researching to collecting data to drawing conclusions.

Communicate

The communication of ideas is an important part of the work of scientists. A discovery that is not reported will not advance the scientific community's understanding or knowledge. Communication among scientists also is important as a way of improving their investigations.

Scientists communicate in many ways, from writing articles in journals and magazines that explain their investigations and experiments, to announcing important discoveries on television and radio. Scientists also share ideas with colleagues on the Internet or present them as lectures, like the student is doing in **Figure 15.**

Figure 15 A student communicates to his peers about his investigation.

SAFETY SYMBOLS

	HAZARD	EXAMPLES	PRECAUTION	REMEDY
DISPOSAL	Special disposal procedures need to be followed.	certain chemicals, living organisms	Do not dispose of these materials in the sink or trash can.	Dispose of wastes as directed by your teacher.
BIOLOGICAL	Organisms or other biological materials that might be harmful to humans	bacteria, fungi, blood, unpreserved tissues, plant materials	Avoid skin contact with these materials. Wear mask or gloves.	Notify your teacher if you suspect contact with material. Wash hands thoroughly.
EXTREME TEMPERATURE	Objects that can burn skin by being too cold or too hot	boiling liquids, hot plates, dry ice, liquid nitrogen	Use proper protection when handling.	Go to your teacher for first aid.
SHARP OBJECT	Use of tools or glassware that can easily puncture or slice skin	razor blades, pins, scalpels, pointed tools, dissecting probes, broken glass	Practice commonsense behavior and follow guidelines for use of the tool.	Go to your teacher for first aid.
FUME	Possible danger to respiratory tract from fumes	ammonia, acetone, nail polish remover, heated sulfur, moth balls	Make sure there is good ventilation. Never smell fumes directly. Wear a mask.	Leave foul area and notify your teacher immediately.
ELECTRICAL	Possible danger from electrical shock or burn	improper grounding, liquid spills, short circuits, exposed wires	Double-check setup with teacher. Check condition of wires and apparatus.	Do not attempt to fix electrical problems. Notify your teacher immediately.
IRRITANT	Substances that can irritate the skin or mucous membranes of the respiratory tract	pollen, moth balls, steel wool, fiberglass, potassium permanganate	Wear dust mask and gloves. Practice extra care when handling these materials.	Go to your teacher for first aid.
CHEMICAL	Chemicals can react with and destroy tissue and other materials	bleaches such as hydrogen peroxide; acids such as sulfuric acid, hydrochloric acid; bases such as ammonia, sodium hydroxide	Wear goggles, gloves, and an apron.	Immediately flush the affected area with water and notify your teacher.
TOXIC	Substance may be poisonous if touched, inhaled, or swallowed.	mercury, many metal compounds, iodine, poinsettia plant parts	Follow your teacher's instructions.	Always wash hands thoroughly after use. Go to your teacher for first aid.
FLAMMABLE	Flammable chemicals may be ignited by open flame, spark, or exposed heat.	alcohol, kerosene, potassium permanganate	Avoid open flames and heat when using flammable chemicals.	Notify your teacher immediately. Use fire safety equipment if applicable.
OPEN FLAME	Open flame in use, may cause fire.	hair, clothing, paper, synthetic materials	Tie back hair and loose clothing. Follow teacher's instruction on lighting and extinguishing flames.	Notify your teacher immediately. Use fire safety equipment if applicable.

 Eye Safety Proper eye protection should be worn at all times by anyone performing or observing science activities.

 Clothing Protection This symbol appears when substances could stain or burn clothing.

 Animal Safety This symbol appears when safety of animals and students must be ensured.

 Handwashing After the lab, wash hands with soap and water before removing goggles.

Safety in the Science Laboratory

The science laboratory is a safe place to work if you follow standard safety procedures. Being responsible for your own safety helps to make the entire laboratory a safer place for everyone. When performing any lab, read and apply the caution statements and safety symbol listed at the beginning of the lab.

General Safety Rules

1. Obtain your teacher's permission to begin all investigations and use laboratory equipment.

2. Study the procedure. Ask your teacher any questions. Be sure you understand safety symbols shown on the page.

3. Notify your teacher about allergies or other health conditions which can affect your participation in a lab.

4. Learn and follow use and safety procedures for your equipment. If unsure, ask your teacher.

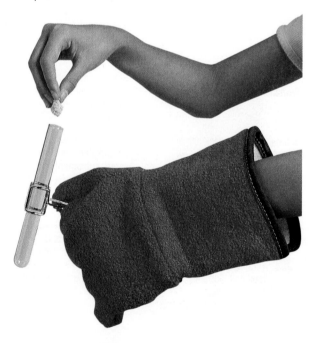

5. Never eat, drink, chew gum, apply cosmetics, or do any personal grooming in the lab. Never use lab glassware as food or drink containers. Keep your hands away from your face and mouth.

6. Know the location and proper use of the safety shower, eye wash, fire blanket, and fire alarm.

Prevent Accidents

1. Use the safety equipment provided to you. Goggles and a safety apron should be worn during investigations.

2. Do NOT use hair spray, mousse, or other flammable hair products. Tie back long hair and tie down loose clothing.

3. Do NOT wear sandals or other open-toed shoes in the lab.

4. Remove jewelry on hands and wrists. Loose jewelry, such as chains and long necklaces, should be removed to prevent them from getting caught in equipment.

5. Do not taste any substances or draw any material into a tube with your mouth.

6. Proper behavior is expected in the lab. Practical jokes and fooling around can lead to accidents and injury.

7. Keep your work area uncluttered.

Laboratory Work

1. Collect and carry all equipment and materials to your work area before beginning a lab.

2. Remain in your own work area unless given permission by your teacher to leave it.

3. Dispose of chemicals and other materials as directed by your teacher. Place broken glass and solid substances in the proper containers. Never discard materials in the sink.

4. Clean your work area.

5. Wash your hands with soap and water thoroughly BEFORE removing your goggles.

Emergencies

1. Report any fire, electrical shock, glassware breakage, spill, or injury, no matter how small, to your teacher immediately. Follow his or her instructions.

2. If your clothing should catch fire, STOP, DROP, and ROLL. If possible, smother it with the fire blanket or get under a safety shower. NEVER RUN.

3. If a fire should occur, turn off all gas and leave the room according to established procedures.

4. In most instances, your teacher will clean up spills. Do NOT attempt to clean up spills unless you are given permission and instructions to do so.

5. If chemicals come into contact with your eyes or skin, notify your teacher immediately. Use the eyewash or flush your skin or eyes with large quantities of water.

6. The fire extinguisher and first-aid kit should only be used by your teacher unless it is an extreme emergency and you have been given permission.

7. If someone is injured or becomes ill, only a professional medical provider or someone certified in first aid should perform first-aid procedures.

3. Always slant test tubes away from yourself and others when heating them, adding substances to them, or rinsing them.

4. If instructed to smell a substance in a container, hold the container a short distance away and fan vapors towards your nose.

5. Do NOT substitute other chemicals/substances for those in the materials list unless instructed to do so by your teacher.

6. Do NOT take any materials or chemicals outside of the laboratory.

7. Stay out of storage areas unless instructed to be there and supervised by your teacher.

Laboratory Cleanup

1. Turn off all burners, water, and gas, and disconnect all electrical devices.

2. Clean all pieces of equipment and return all materials to their proper places.

EXTRA  Labs

From Your Kitchen, Junk Drawer, or Yard

1 Exploding Bag

▶ *Real-World Question*

What happens when a bag pops?

Possible Materials

• paper bag or plastic produce bag

▶ *Procedure*

1. Obtain a paper lunch bag. Smooth out the bag on a flat surface if it has any wrinkles.
2. Hold the neck of the bag and blow air into it until it is completely filled. The sides of the bag should be stretched out completely.
3. Twist the neck of the bag tightly to prevent air from escaping.
4. Pop the paper bag between your palms and observe what happens.
5. Examine the bag after you pop it. Observe any changes in the bag.

▶ *Conclude and Apply*

1. Describe what happened when you popped the bag.
2. Infer why this happened to the bag.

2 Seeing Sound

▶ *Real-World Question*

Is it possible to see sound waves?

Possible Materials

• scissors
• rubber band
• twigs, curved and stiff
• sewing thread
• puffed-rice cereal
• clothing hanger

▶ *Procedure*

1. Make a rubber-band bow as follows: Cut the rubber band at one point. Tie the rubber band around opposite ends of the twig so that it looks like an archery bow. Make sure the rubber band is tight like a guitar string.
2. Cut about 10 pieces of thread to equal lengths, 10 to 15 cm long.
3. With each piece of thread, tie one end around a kernel of puffed rice and the other end around the bottom of a clothing hanger. Space the hanging kernels about 1 cm apart.
4. Hook the clothing hanger over something so that the threads and cereal hang freely.
5. Hold the rubber-band bow so that the rubber band is just underneath the central hanging kernels. Pluck the rubber band, being careful not to touch the kernels. Write down your observations.

▶ *Conclude and Apply*

1. Explain how this experiment relates to the fact that sound travels through air using compression waves.

Adult supervision required for all labs.

Extra Try at Home Labs

3 Black Light

▶ Real-World Question
How do you know that ultraviolet waves exist?

Possible Materials
- normal lamp with a white lightbulb
- ultraviolet light source (a black light)
- white paper
- laundry detergent
- glow-in-the-dark plastic toy
- a variety of rocks and minerals
- a variety of flowers and plants
- a variety of household cleaners
- different colors and materials of clothing

▶ Procedure
1. Dab a small amount of laundry detergent on some white paper and place it somewhere to dry.
2. Place the black light in a dark room and turn it on.
3. Write down what the detergent looks like under a normal lamp. Then, place the paper under the black light. Write down what you see.
4. Place other items under the different lights and write down what you see.

▶ Conclude and Apply
1. Describe the difference between the way things look under normal light and the way they look under ultraviolet light.
2. Explain how you know from this experiment that ultraviolet waves exist.

4 Light in Liquids

▶ Real-World Question
What happens to light when it passes through different liquids found in your kitchen?

Possible Materials
- flashlight
- glass
- orange juice
- water
- milk
- maple syrup
- white vinegar
- red vinegar
- honey
- molasses
- milk
- fruit juice
- powdered drink mix
- salad dressing
- salsa

▶ Procedure
1. Fill a glass with water and darken the room. Shine the beam of a flashlight through the glass and observe how much of the light passes through the water.
2. Identify water as an opaque, translucent, or transparent substance.
3. Repeat steps 1–2 to test a wide variety of other liquids found in your kitchen. Use the original containers when you can, but remove any labels from containers that block the light beam.

▶ Conclude and Apply
1. Identify all the opaque liquids you tested.
2. Identify all the translucent liquids you tested.
3. Identify all the transparent liquids you tested.

Computer Skills

People who study science rely on computers, like the one in **Figure 16,** to record and store data and to analyze results from investigations. Whether you work in a laboratory or just need to write a lab report with tables, good computer skills are a necessity.

Using the computer comes with responsibility. Issues of ownership, security, and privacy can arise. Remember, if you did not author the information you are using, you must provide a source for your information. Also, anything on a computer can be accessed by others. Do not put anything on the computer that you would not want everyone to know. To add more security to your work, use a password.

Use a Word Processing Program

A computer program that allows you to type your information, change it as many times as you need to, and then print it out is called a word processing program. Word processing programs also can be used to make tables.

Figure 16 A computer will make reports neater and more professional looking.

Learn the Skill To start your word processing program, a blank document, sometimes called "Document 1," appears on the screen. To begin, start typing. To create a new document, click the *New* button on the standard tool bar. These tips will help you format the document.

- The program will automatically move to the next line; press *Enter* if you wish to start a new paragraph.
- Symbols, called non-printing characters, can be hidden by clicking the *Show/Hide* button on your toolbar.
- To insert text, move the cursor to the point where you want the insertion to go, click on the mouse once, and type the text.
- To move several lines of text, select the text and click the *Cut* button on your toolbar. Then position your cursor in the location that you want to move the cut text and click *Paste.* If you move to the wrong place, click *Undo.*
- The spell check feature does not catch words that are misspelled to look like other words, like "cold" instead of "gold." Always reread your document to catch all spelling mistakes.
- To learn about other word processing methods, read the user's manual or click on the *Help* button.
- You can integrate databases, graphics, and spreadsheets into documents by copying from another program and pasting it into your document, or by using desktop publishing (DTP). DTP software allows you to put text and graphics together to finish your document with a professional look. This software varies in how it is used and its capabilities.

Use a Database

A collection of facts stored in a computer and sorted into different fields is called a database. A database can be reorganized in any way that suits your needs.

Learn the Skill A computer program that allows you to create your own database is a database management system (DBMS). It allows you to add, delete, or change information. Take time to get to know the features of your database software.

- Determine what facts you would like to include and research to collect your information.
- Determine how you want to organize the information.
- Follow the instructions for your particular DBMS to set up fields. Then enter each item of data in the appropriate field.
- Follow the instructions to sort the information in order of importance.
- Evaluate the information in your database, and add, delete, or change as necessary.

Use the Internet

The Internet is a global network of computers where information is stored and shared. To use the Internet, like the students in **Figure 17,** you need a modem to connect your computer to a phone line and an Internet Service Provider account.

Learn the Skill To access internet sites and information, use a "Web browser," which lets you view and explore pages on the World Wide Web. Each page is its own site, and each site has its own address, called a URL. Once you have found a Web browser, follow these steps for a search (this also is how you search a database).

Figure 17 The Internet allows you to search a global network for a variety of information.

- Be as specific as possible. If you know you want to research "gold," don't type in "elements." Keep narrowing your search until you find what you want.
- Web sites that end in *.com* are commercial Web sites; *.org, .edu,* and *.gov* are non-profit, educational, or government Web sites.
- Electronic encyclopedias, almanacs, indexes, and catalogs will help locate and select relevant information.
- Develop a "home page" with relative ease. When developing a Web site, NEVER post pictures or disclose personal information such as location, names, or phone numbers. Your school or community usually can host your Web site. A basic understanding of HTML (hypertext mark-up language), the language of Web sites, is necessary. Software that creates HTML code is called authoring software, and can be downloaded free from many Web sites. This software allows text and pictures to be arranged as the software is writing the HTML code.

Use a Spreadsheet

A spreadsheet, shown in **Figure 18,** can perform mathematical functions with any data arranged in columns and rows. By entering a simple equation into a cell, the program can perform operations in specific cells, rows, or columns.

Learn the Skill Each column (vertical) is assigned a letter, and each row (horizontal) is assigned a number. Each point where a row and column intersect is called a cell, and is labeled according to where it is located—Column A, Row 1 (A1).

- Decide how to organize the data, and enter it in the correct row or column.
- Spreadsheets can use standard formulas or formulas can be customized to calculate cells.
- To make a change, click on a cell to make it activate, and enter the edited data or formula.
- Spreadsheets also can display your results in graphs. Choose the style of graph that best represents the data.

	A	B	C	D	E
1	Test Runs	Time	Distance	Speed	
2	Car 1	5 mins	5 miles	60 mph	
3	Car 2	10 mins	4 miles	24 mph	
4	Car 3	6 mins	3 miles	30 mph	

Figure 18 A spreadsheet allows you to perform mathematical operations on your data.

Use Graphics Software

Adding pictures, called graphics, to your documents is one way to make your documents more meaningful and exciting. This software adds, edits, and even constructs graphics. There is a variety of graphics software programs. The tools used for drawing can be a mouse, keyboard, or other specialized devices. Some graphics programs are simple. Others are complicated, called computer-aided design (CAD) software.

Learn the Skill It is important to have an understanding of the graphics software being used before starting. The better the software is understood, the better the results. The graphics can be placed in a word-processing document.

- Clip art can be found on a variety of internet sites, and on CDs. These images can be copied and pasted into your document.
- When beginning, try editing existing drawings, then work up to creating drawings.
- The images are made of tiny rectangles of color called pixels. Each pixel can be altered.
- Digital photography is another way to add images. The photographs in the memory of a digital camera can be downloaded into a computer, then edited and added to the document.
- Graphics software also can allow animation. The software allows drawings to have the appearance of movement by connecting basic drawings automatically. This is called in-betweening, or tweening.
- Remember to save often.

Presentation Skills

Develop Multimedia Presentations

Most presentations are more dynamic if they include diagrams, photographs, videos, or sound recordings, like the one shown in **Figure 19.** A multimedia presentation involves using stereos, overhead projectors, televisions, computers, and more.

Learn the Skill Decide the main points of your presentation, and what types of media would best illustrate those points.

- Make sure you know how to use the equipment you are working with.
- Practice the presentation using the equipment several times.
- Enlist the help of a classmate to push play or turn lights out for you. Be sure to practice your presentation with him or her.
- If possible, set up all of the equipment ahead of time, and make sure everything is working properly.

Figure 19 These students are engaging the audience using a variety of tools.

Computer Presentations

There are many different interactive computer programs that you can use to enhance your presentation. Most computers have a compact disc (CD) drive that can play both CDs and digital video discs (DVDs). Also, there is hardware to connect a regular CD, DVD, or VCR. These tools will enhance your presentation.

Another method of using the computer to aid in your presentation is to develop a slide show using a computer program. This can allow movement of visuals at the presenter's pace, and can allow for visuals to build on one another.

Learn the Skill In order to create multimedia presentations on a computer, you need to have certain tools. These may include traditional graphic tools and drawing programs, animation programs, and authoring systems that tie everything together. Your computer will tell you which tools it supports. The most important step is to learn about the tools that you will be using.

- Often, color and strong images will convey a point better than words alone. Use the best methods available to convey your point.
- As with other presentations, practice many times.
- Practice your presentation with the tools you and any assistants will be using.
- Maintain eye contact with the audience. The purpose of using the computer is not to prompt the presenter, but to help the audience understand the points of the presentation.

Math Review

Use Fractions

A fraction compares a part to a whole. In the fraction $\frac{2}{3}$, the 2 represents the part and is the numerator. The 3 represents the whole and is the denominator.

Reduce Fractions To reduce a fraction, you must find the largest factor that is common to both the numerator and the denominator, the greatest common factor (GCF). Divide both numbers by the GCF. The fraction has then been reduced, or it is in its simplest form.

Example Twelve of the 20 chemicals in the science lab are in powder form. What fraction of the chemicals used in the lab are in powder form?

Step 1 Write the fraction.

$$\frac{\text{part}}{\text{whole}} = \frac{12}{20}$$

Step 2 To find the GCF of the numerator and denominator, list all of the factors of each number.

Factors of 12: 1, 2, 3, 4, 6, 12 (the numbers that divide evenly into 12)

Factors of 20: 1, 2, 4, 5, 10, 20 (the numbers that divide evenly into 20)

Step 3 List the common factors.

1, 2, 4.

Step 4 Choose the greatest factor in the list.

The GCF of 12 and 20 is 4.

Step 5 Divide the numerator and denominator by the GCF.

$$\frac{12 \div 4}{20 \div 4} = \frac{3}{5}$$

In the lab, $\frac{3}{5}$ of the chemicals are in powder form.

Practice Problem At an amusement park, 66 of 90 rides have a height restriction. What fraction of the rides, in its simplest form, has a height restriction?

Add and Subtract Fractions To add or subtract fractions with the same denominator, add or subtract the numerators and write the sum or difference over the denominator. After finding the sum or difference, find the simplest form for your fraction.

Example 1 In the forest outside your house, $\frac{1}{8}$ of the animals are rabbits, $\frac{3}{8}$ are squirrels, and the remainder are birds and insects. How many are mammals?

Step 1 Add the numerators.

$$\frac{1}{8} + \frac{3}{8} = \frac{(1 + 3)}{8} = \frac{4}{8}$$

Step 2 Find the GCF.

$$\frac{4}{8} \text{ (GCF, 4)}$$

Step 3 Divide the numerator and denominator by the GCF.

$$\frac{4}{4} = 1, \ \frac{8}{4} = 2$$

$\frac{1}{2}$ of the animals are mammals.

Example 2 If $\frac{7}{16}$ of the Earth is covered by freshwater, and $\frac{1}{16}$ of that is in glaciers, how much freshwater is not frozen?

Step 1 Subtract the numerators.

$$\frac{7}{16} - \frac{1}{16} = \frac{(7 - 1)}{16} = \frac{6}{16}$$

Step 2 Find the GCF.

$$\frac{6}{16} \text{ (GCF, 2)}$$

Step 3 Divide the numerator and denominator by the GCF.

$$\frac{6}{2} = 3, \ \frac{16}{2} = 8$$

$\frac{3}{8}$ of the freshwater is not frozen.

Practice Problem A bicycle rider is going 15 km/h for $\frac{4}{9}$ of his ride, 10 km/h for $\frac{2}{9}$ of his ride, and 8 km/h for the remainder of the ride. How much of his ride is he going over 8 km/h?

Unlike Denominators To add or subtract fractions with unlike denominators, first find the least common denominator (LCD). This is the smallest number that is a common multiple of both denominators. Rename each fraction with the LCD, and then add or subtract. Find the simplest form if necessary.

Example 1 A chemist makes a paste that is $\frac{1}{2}$ table salt (NaCl), $\frac{1}{3}$ sugar ($C_6H_{12}O_6$), and the rest water (H_2O). How much of the paste is a solid?

Step 1 Find the LCD of the fractions.

$\frac{1}{2} + \frac{1}{3}$ (LCD, 6)

Step 2 Rename each numerator and each denominator with the LCD.

$1 \times 3 = 3, \quad 2 \times 3 = 6$
$1 \times 2 = 2, \quad 3 \times 2 = 6$

Step 3 Add the numerators.

$\frac{3}{6} + \frac{2}{6} = \frac{(3 + 2)}{6} = \frac{5}{6}$

$\frac{5}{6}$ of the paste is a solid.

Example 2 The average precipitation in Grand Junction, CO, is $\frac{7}{10}$ inch in November, and $\frac{3}{5}$ inch in December. What is the total average precipitation?

Step 1 Find the LCD of the fractions.

$\frac{7}{10} + \frac{3}{5}$ (LCD, 10)

Step 2 Rename each numerator and each denominator with the LCD.

$7 \times 1 = 7, \quad 10 \times 1 = 10$
$3 \times 2 = 6, \quad 5 \times 2 = 10$

Step 3 Add the numerators.

$\frac{7}{10} + \frac{6}{10} = \frac{(7 + 6)}{10} = \frac{13}{10}$

$\frac{13}{10}$ inches total precipitation, or $1\frac{3}{10}$ inches.

Practice Problem On an electric bill, about $\frac{1}{8}$ of the energy is from solar energy and about $\frac{1}{10}$ is from wind power. How much of the total bill is from solar energy and wind power combined?

Example 3 In your body, $\frac{7}{10}$ of your muscle contractions are involuntary (cardiac and smooth muscle tissue). Smooth muscle makes $\frac{3}{15}$ of your muscle contractions. How many of your muscle contractions are made by cardiac muscle?

Step 1 Find the LCD of the fractions.

$\frac{7}{10} - \frac{3}{15}$ (LCD, 30)

Step 2 Rename each numerator and each denominator with the LCD.

$7 \times 3 = 21, \quad 10 \times 3 = 30$
$3 \times 2 = 6, \quad 15 \times 2 = 30$

Step 3 Subtract the numerators.

$\frac{21}{30} - \frac{6}{30} = \frac{(21 - 6)}{30} = \frac{15}{30}$

Step 4 Find the GCF.

$\frac{15}{30}$ (GCF, 15)

$\frac{1}{2}$

$\frac{1}{2}$ of all muscle contractions are cardiac muscle.

Example 4 Tony wants to make cookies that call for $\frac{3}{4}$ of a cup of flour, but he only has $\frac{1}{3}$ of a cup. How much more flour does he need?

Step 1 Find the LCD of the fractions.

$\frac{3}{4} - \frac{1}{3}$ (LCD, 12)

Step 2 Rename each numerator and each denominator with the LCD.

$3 \times 3 = 9, \quad 4 \times 3 = 12$
$1 \times 4 = 4, \quad 3 \times 4 = 12$

Step 3 Subtract the numerators.

$\frac{9}{12} - \frac{4}{12} = \frac{(9 - 4)}{12} = \frac{5}{12}$

$\frac{5}{12}$ of a cup of flour.

Practice Problem Using the information provided to you in Example 3 above, determine how many muscle contractions are voluntary (skeletal muscle).

Math Skill Handbook

Multiply Fractions To multiply with fractions, multiply the numerators and multiply the denominators. Find the simplest form if necessary.

Example Multiply $\frac{3}{5}$ by $\frac{1}{3}$.

Step 1 Multiply the numerators and denominators.

$$\frac{3}{5} \times \frac{1}{3} = \frac{(3 \times 1)}{(5 \times 3)} = \frac{3}{15}$$

Step 2 Find the GCF.

$$\frac{3}{15} \quad (\text{GCF, 3})$$

Step 3 Divide the numerator and denominator by the GCF.

$$\frac{3}{3} = 1, \quad \frac{15}{3} = 5$$

$$\frac{1}{5}$$

$\frac{3}{5}$ multiplied by $\frac{1}{3}$ is $\frac{1}{5}$.

Practice Problem Multiply $\frac{3}{14}$ by $\frac{5}{16}$.

Find a Reciprocal Two numbers whose product is 1 are called multiplicative inverses, or reciprocals.

Example Find the reciprocal of $\frac{3}{8}$.

Step 1 Inverse the fraction by putting the denominator on top and the numerator on the bottom.

$$\frac{8}{3}$$

The reciprocal of $\frac{3}{8}$ is $\frac{8}{3}$.

Practice Problem Find the reciprocal of $\frac{4}{9}$.

Divide Fractions To divide one fraction by another fraction, multiply the dividend by the reciprocal of the divisor. Find the simplest form if necessary.

Example 1 Divide $\frac{1}{9}$ by $\frac{1}{3}$.

Step 1 Find the reciprocal of the divisor.

The reciprocal of $\frac{1}{3}$ is $\frac{3}{1}$.

Step 2 Multiply the dividend by the reciprocal of the divisor.

$$\frac{\frac{1}{9}}{\frac{1}{3}} = \frac{1}{9} \times \frac{3}{1} = \frac{(1 \times 3)}{(9 \times 1)} = \frac{3}{9}$$

Step 3 Find the GCF.

$$\frac{3}{9} \quad (\text{GCF, 3})$$

Step 4 Divide the numerator and denominator by the GCF.

$$\frac{3}{3} = 1, \quad \frac{9}{3} = 3$$

$$\frac{1}{3}$$

$\frac{1}{9}$ divided by $\frac{1}{3}$ is $\frac{1}{3}$.

Example 2 Divide $\frac{3}{5}$ by $\frac{1}{4}$.

Step 1 Find the reciprocal of the divisor.

The reciprocal of $\frac{1}{4}$ is $\frac{4}{1}$.

Step 2 Multiply the dividend by the reciprocal of the divisor.

$$\frac{\frac{3}{5}}{\frac{1}{4}} = \frac{3}{5} \times \frac{4}{1} = \frac{(3 \times 4)}{(5 \times 1)} = \frac{12}{5}$$

$\frac{3}{5}$ divided by $\frac{1}{4}$ is $\frac{12}{5}$ or $2\frac{2}{5}$.

Practice Problem Divide $\frac{3}{11}$ by $\frac{7}{10}$.

Math Skill Handbook

Use Ratios

When you compare two numbers by division, you are using a ratio. Ratios can be written 3 to 5, 3:5, or $\frac{3}{5}$. Ratios, like fractions, also can be written in simplest form.

Ratios can represent probabilities, also called odds. This is a ratio that compares the number of ways a certain outcome occurs to the number of outcomes. For example, if you flip a coin 100 times, what are the odds that it will come up heads? There are two possible outcomes, heads or tails, so the odds of coming up heads are 50:100. Another way to say this is that 50 out of 100 times the coin will come up heads. In its simplest form, the ratio is 1:2.

Example 1 A chemical solution contains 40 g of salt and 64 g of baking soda. What is the ratio of salt to baking soda as a fraction in simplest form?

Step 1 Write the ratio as a fraction.
$$\frac{salt}{baking\ soda} = \frac{40}{64}$$

Step 2 Express the fraction in simplest form.
The GCF of 40 and 64 is 8.
$$\frac{40}{64} = \frac{40 \div 8}{64 \div 8} = \frac{5}{8}$$

The ratio of salt to baking soda in the sample is 5:8.

Example 2 Sean rolls a 6-sided die 6 times. What are the odds that the side with a 3 will show?

Step 1 Write the ratio as a fraction.
$$\frac{number\ of\ sides\ with\ a\ 3}{number\ of\ sides} = \frac{1}{6}$$

Step 2 Multiply by the number of attempts.
$$\frac{1}{6} \times 6\ attempts = \frac{6}{6}\ attempts = 1\ attempt$$

1 attempt out of 6 will show a 3.

Practice Problem Two metal rods measure 100 cm and 144 cm in length. What is the ratio of their lengths in simplest form?

Use Decimals

A fraction with a denominator that is a power of ten can be written as a decimal. For example, 0.27 means $\frac{27}{100}$. The decimal point separates the ones place from the tenths place.

Any fraction can be written as a decimal using division. For example, the fraction $\frac{5}{8}$ can be written as a decimal by dividing 5 by 8. Written as a decimal, it is 0.625.

Add or Subtract Decimals When adding and subtracting decimals, line up the decimal points before carrying out the operation.

Example 1 Find the sum of 47.68 and 7.80.

Step 1 Line up the decimal places when you write the numbers.
```
  47.68
+  7.80
```

Step 2 Add the decimals.
```
  47.68
+  7.80
  55.48
```

The sum of 47.68 and 7.80 is 55.48.

Example 2 Find the difference of 42.17 and 15.85.

Step 1 Line up the decimal places when you write the number.
```
  42.17
- 15.85
```

Step 2 Subtract the decimals.
```
  42.17
- 15.85
  26.32
```

The difference of 42.17 and 15.85 is 26.32.

Practice Problem Find the sum of 1.245 and 3.842.

Multiply Decimals To multiply decimals, multiply the numbers like any other number, ignoring the decimal point. Count the decimal places in each factor. The product will have the same number of decimal places as the sum of the decimal places in the factors.

Example Multiply 2.4 by 5.9.

Step 1 Multiply the factors like two whole numbers.
$$24 \times 59 = 1416$$

Step 2 Find the sum of the number of decimal places in the factors. Each factor has one decimal place, for a sum of two decimal places.

Step 3 The product will have two decimal places.
14.16

The product of 2.4 and 5.9 is 14.16.

Practice Problem Multiply 4.6 by 2.2.

Divide Decimals When dividing decimals, change the divisor to a whole number. To do this, multiply both the divisor and the dividend by the same power of ten. Then place the decimal point in the quotient directly above the decimal point in the dividend. Then divide as you do with whole numbers.

Example Divide 8.84 by 3.4.

Step 1 Multiply both factors by 10.
$$3.4 \times 10 = 34, \ 8.84 \times 10 = 88.4$$

Step 2 Divide 88.4 by 34.

```
        2.6
   34)88.4
     −68
      204
     −204
        0
```

8.84 divided by 3.4 is 2.6.

Practice Problem Divide 75.6 by 3.6.

Use Proportions

An equation that shows that two ratios are equivalent is a proportion. The ratios $\frac{2}{4}$ and $\frac{5}{10}$ are equivalent, so they can be written as $\frac{2}{4} = \frac{5}{10}$. This equation is a proportion.

When two ratios form a proportion, the cross products are equal. To find the cross products in the proportion $\frac{2}{4} = \frac{5}{10}$, multiply the 2 and the 10, and the 4 and the 5. Therefore $2 \times 10 = 4 \times 5$, or $20 = 20$.

Because you know that both proportions are equal, you can use cross products to find a missing term in a proportion. This is known as solving the proportion.

Example The heights of a tree and a pole are proportional to the lengths of their shadows. The tree casts a shadow of 24 m when a 6-m pole casts a shadow of 4 m. What is the height of the tree?

Step 1 Write a proportion.
$$\frac{\text{height of tree}}{\text{height of pole}} = \frac{\text{length of tree's shadow}}{\text{length of pole's shadow}}$$

Step 2 Substitute the known values into the proportion. Let h represent the unknown value, the height of the tree.
$$\frac{h}{6} = \frac{24}{4}$$

Step 3 Find the cross products.
$$h \times 4 = 6 \times 24$$

Step 4 Simplify the equation.
$$4h = 144$$

Step 5 Divide each side by 4.
$$\frac{4h}{4} = \frac{144}{4}$$
$$h = 36$$

The height of the tree is 36 m.

Practice Problem The ratios of the weights of two objects on the Moon and on Earth are in proportion. A rock weighing 3 N on the Moon weighs 18 N on Earth. How much would a rock that weighs 5 N on the Moon weigh on Earth?

Use Percentages

The word *percent* means "out of one hundred." It is a ratio that compares a number to 100. Suppose you read that 77 percent of the Earth's surface is covered by water. That is the same as reading that the fraction of the Earth's surface covered by water is $\frac{77}{100}$. To express a fraction as a percent, first find the equivalent decimal for the fraction. Then, multiply the decimal by 100 and add the percent symbol.

Example Express $\frac{13}{20}$ as a percent.

Step 1 Find the equivalent decimal for the fraction.

$$\begin{array}{r} 0.65 \\ 20)\overline{13.00} \\ \underline{12\,0} \\ 1\,00 \\ \underline{1\,00} \\ 0 \end{array}$$

Step 2 Rewrite the fraction $\frac{13}{20}$ as 0.65.

Step 3 Multiply 0.65 by 100 and add the % sign.
$$0.65 \times 100 = 65 = 65\%$$

So, $\frac{13}{20} = 65\%$.

This also can be solved as a proportion.

Example Express $\frac{13}{20}$ as a percent.

Step 1 Write a proportion.
$$\frac{13}{20} = \frac{x}{100}$$

Step 2 Find the cross products.
$$1300 = 20x$$

Step 3 Divide each side by 20.
$$\frac{1300}{20} = \frac{20x}{20}$$
$$65\% = x$$

Practice Problem In one year, 73 of 365 days were rainy in one city. What percent of the days in that city were rainy?

Solve One-Step Equations

A statement that two things are equal is an equation. For example, $A = B$ is an equation that states that A is equal to B.

An equation is solved when a variable is replaced with a value that makes both sides of the equation equal. To make both sides equal the inverse operation is used. Addition and subtraction are inverses, and multiplication and division are inverses.

Example 1 Solve the equation $x - 10 = 35$.

Step 1 Find the solution by adding 10 to each side of the equation.
$$x - 10 = 35$$
$$x - 10 + 10 = 35 + 10$$
$$x = 45$$

Step 2 Check the solution.
$$x - 10 = 35$$
$$45 - 10 = 35$$
$$35 = 35$$

Both sides of the equation are equal, so $x = 45$.

Example 2 In the formula $a = bc$, find the value of c if $a = 20$ and $b = 2$.

Step 1 Rearrange the formula so the unknown value is by itself on one side of the equation by dividing both sides by b.
$$a = bc$$
$$\frac{a}{b} = \frac{bc}{b}$$
$$\frac{a}{b} = c$$

Step 2 Replace the variables a and b with the values that are given.
$$\frac{a}{b} = c$$
$$\frac{20}{2} = c$$
$$10 = c$$

Step 3 Check the solution.
$$a = bc$$
$$20 = 2 \times 10$$
$$20 = 20$$

Both sides of the equation are equal, so $c = 10$ is the solution when $a = 20$ and $b = 2$.

Practice Problem In the formula $h = gd$, find the value of d if $g = 12.3$ and $h = 17.4$.

Use Statistics

The branch of mathematics that deals with collecting, analyzing, and presenting data is statistics. In statistics, there are three common ways to summarize data with a single number—the mean, the median, and the mode.

The **mean** of a set of data is the arithmetic average. It is found by adding the numbers in the data set and dividing by the number of items in the set.

The **median** is the middle number in a set of data when the data are arranged in numerical order. If there were an even number of data points, the median would be the mean of the two middle numbers.

The **mode** of a set of data is the number or item that appears most often.

Another number that often is used to describe a set of data is the range. The **range** is the difference between the largest number and the smallest number in a set of data.

A **frequency table** shows how many times each piece of data occurs, usually in a survey. **Table 2** below shows the results of a student survey on favorite color.

Table 2 Student Color Choice		
Color	**Tally**	**Frequency**
red	||||	4
blue	︴︲︲	5
black	||	2
green	|||	3
purple	︴︲︲ ||	7
yellow	︴︲︲ |	6

Based on the frequency table data, which color is the favorite?

Example The speeds (in m/s) for a race car during five different time trials are 39, 37, 44, 36, and 44.

To find the mean:

Step 1 Find the sum of the numbers.
$$39 + 37 + 44 + 36 + 44 = 200$$

Step 2 Divide the sum by the number of items, which is 5.
$$200 \div 5 = 40$$

The mean is 40 m/s.

To find the median:

Step 1 Arrange the measures from least to greatest.
36, 37, 39, 44, 44

Step 2 Determine the middle measure.
36, 37, <u>39</u>, 44, 44

The median is 39 m/s.

To find the mode:

Step 1 Group the numbers that are the same together.
44, 44, 36, 37, 39

Step 2 Determine the number that occurs most in the set.
<u>44, 44</u>, 36, 37, 39

The mode is 44 m/s.

To find the range:

Step 1 Arrange the measures from largest to smallest.
44, 44, 39, 37, 36

Step 2 Determine the largest and smallest measures in the set.
<u>44</u>, 44, 39, 37, <u>36</u>

Step 3 Find the difference between the largest and smallest measures.
$$44 - 36 = 8$$

The range is 8 m/s.

Practice Problem Find the mean, median, mode, and range for the data set 8, 4, 12, 8, 11, 14, 16.

Use Geometry

The branch of mathematics that deals with the measurement, properties, and relationships of points, lines, angles, surfaces, and solids is called geometry.

Perimeter The **perimeter** (P) is the distance around a geometric figure. To find the perimeter of a rectangle, add the length and width and multiply that sum by two, or $2(l + w)$. To find perimeters of irregular figures, add the length of the sides.

Example 1 Find the perimeter of a rectangle that is 3 m long and 5 m wide.

Step 1 You know that the perimeter is 2 times the sum of the width and length.
$$P = 2(3\text{ m} + 5\text{ m})$$

Step 2 Find the sum of the width and length.
$$P = 2(8\text{ m})$$

Step 3 Multiply by 2.
$$P = 16\text{ m}$$

The perimeter is 16 m.

Example 2 Find the perimeter of a shape with sides measuring 2 cm, 5 cm, 6 cm, 3 cm.

Step 1 You know that the perimeter is the sum of all the sides.
$$P = 2 + 5 + 6 + 3$$

Step 2 Find the sum of the sides.
$$P = 2 + 5 + 6 + 3$$
$$P = 16$$

The perimeter is 16 cm.

Practice Problem Find the perimeter of a rectangle with a length of 18 m and a width of 7 m.

Practice Problem Find the perimeter of a triangle measuring 1.6 cm by 2.4 cm by 2.4 cm.

Area of a Rectangle The **area** (A) is the number of square units needed to cover a surface. To find the area of a rectangle, multiply the length times the width, or $l \times w$. When finding area, the units also are multiplied. Area is given in square units.

Example Find the area of a rectangle with a length of 1 cm and a width of 10 cm.

Step 1 You know that the area is the length multiplied by the width.
$$A = (1\text{ cm} \times 10\text{ cm})$$

Step 2 Multiply the length by the width. Also multiply the units.
$$A = 10\text{ cm}^2$$

The area is 10 cm².

Practice Problem Find the area of a square whose sides measure 4 m.

Area of a Triangle To find the area of a triangle, use the formula:

$$A = \frac{1}{2}(\text{base} \times \text{height})$$

The base of a triangle can be any of its sides. The height is the perpendicular distance from a base to the opposite endpoint, or vertex.

Example Find the area of a triangle with a base of 18 m and a height of 7 m.

Step 1 You know that the area is $\frac{1}{2}$ the base times the height.
$$A = \frac{1}{2}(18\text{ m} \times 7\text{ m})$$

Step 2 Multiply $\frac{1}{2}$ by the product of 18 × 7. Multiply the units.
$$A = \frac{1}{2}(126\text{ m}^2)$$
$$A = 63\text{ m}^2$$

The area is 63 m².

Practice Problem Find the area of a triangle with a base of 27 cm and a height of 17 cm.

Circumference of a Circle The **diameter** (d) of a circle is the distance across the circle through its center, and the **radius** (r) is the distance from the center to any point on the circle. The radius is half of the diameter. The distance around the circle is called the **circumference** (C). The formula for finding the circumference is:

$$C = 2\pi r \ or \ C = \pi d$$

The circumference divided by the diameter is always equal to 3.1415926... This nonterminating and nonrepeating number is represented by the Greek letter π (pi). An approximation often used for π is 3.14.

Example 1 Find the circumference of a circle with a radius of 3 m.

Step 1 You know the formula for the circumference is 2 times the radius times π.
$$C = 2\pi(3)$$

Step 2 Multiply 2 times the radius.
$$C = 6\pi$$

Step 3 Multiply by π.
$$C = 19 \ m$$

The circumference is 19 m.

Example 2 Find the circumference of a circle with a diameter of 24.0 cm.

Step 1 You know the formula for the circumference is the diameter times π.
$$C = \pi(24.0)$$

Step 2 Multiply the diameter by π.
$$C = 75.4 \ cm$$

The circumference is 75.4 cm.

Practice Problem Find the circumference of a circle with a radius of 19 cm.

Area of a Circle The formula for the area of a circle is:
$$A = \pi r^2$$

Example 1 Find the area of a circle with a radius of 4.0 cm.

Step 1 $A = \pi(4.0)^2$

Step 2 Find the square of the radius.
$$A = 16\pi$$

Step 3 Multiply the square of the radius by π.
$$A = 50 \ cm^2$$

The area of the circle is 50 cm².

Example 2 Find the area of a circle with a radius of 225 m.

Step 1 $A = \pi(225)^2$

Step 2 Find the square of the radius.
$$A = 50625\pi$$

Step 3 Multiply the square of the radius by π.
$$A = 158962.5$$

The area of the circle is 158,962 m².

Example 3 Find the area of a circle whose diameter is 20.0 mm.

Step 1 You know the formula for the area of a circle is the square of the radius times π, and that the radius is half of the diameter.
$$A = \pi\left(\frac{20.0}{2}\right)^2$$

Step 2 Find the radius.
$$A = \pi(10.0)^2$$

Step 3 Find the square of the radius.
$$A = 100\pi$$

Step 4 Multiply the square of the radius by π.
$$A = 314 \ mm^2$$

The area is 314 mm².

Practice Problem Find the area of a circle with a radius of 16 m.

Volume The measure of space occupied by a solid is the **volume** (V). To find the volume of a rectangular solid multiply the length times width times height, or $V = l \times w \times h$. It is measured in cubic units, such as cubic centimeters (cm^3).

Example Find the volume of a rectangular solid with a length of 2.0 m, a width of 4.0 m, and a height of 3.0 m.

Step 1 You know the formula for volume is the length times the width times the height.
$$V = 2.0 \text{ m} \times 4.0 \text{ m} \times 3.0 \text{ m}$$

Step 2 Multiply the length times the width times the height.
$$V = 24 \text{ m}^3$$

The volume is 24 m^3.

Practice Problem Find the volume of a rectangular solid that is 8 m long, 4 m wide, and 4 m high.

To find the volume of other solids, multiply the area of the base times the height.

Example 1 Find the volume of a solid that has a triangular base with a length of 8.0 m and a height of 7.0 m. The height of the entire solid is 15.0 m.

Step 1 You know that the base is a triangle, and the area of a triangle is $\frac{1}{2}$ the base times the height, and the volume is the area of the base times the height.
$$V = \left[\frac{1}{2} (b \times h) \right] \times 15$$

Step 2 Find the area of the base.
$$V = \left[\frac{1}{2} (8 \times 7) \right] \times 15$$
$$V = \left(\frac{1}{2} \times 56 \right) \times 15$$

Step 3 Multiply the area of the base by the height of the solid.
$$V = 28 \times 15$$
$$V = 420 \text{ m}^3$$

The volume is 420 m^3.

Example 2 Find the volume of a cylinder that has a base with a radius of 12.0 cm, and a height of 21.0 cm.

Step 1 You know that the base is a circle, and the area of a circle is the square of the radius times π, and the volume is the area of the base times the height.
$$V = (\pi r^2) \times 21$$
$$V = (\pi 12^2) \times 21$$

Step 2 Find the area of the base.
$$V = 144\pi \times 21$$
$$V = 452 \times 21$$

Step 3 Multiply the area of the base by the height of the solid.
$$V = 9490 \text{ cm}^3$$

The volume is 9490 cm^3.

Example 3 Find the volume of a cylinder that has a diameter of 15 mm and a height of 4.8 mm.

Step 1 You know that the base is a circle with an area equal to the square of the radius times π. The radius is one-half the diameter. The volume is the area of the base times the height.
$$V = (\pi r^2) \times 4.8$$
$$V = \left[\pi \left(\frac{1}{2} \times 15 \right)^2 \right] \times 4.8$$
$$V = (\pi 7.5^2) \times 4.8$$

Step 2 Find the area of the base.
$$V = 56.25\pi \times 4.8$$
$$V = 176.63 \times 4.8$$

Step 3 Multiply the area of the base by the height of the solid.
$$V = 847.8$$

The volume is 847.8 mm^3.

Practice Problem Find the volume of a cylinder with a diameter of 7 cm in the base and a height of 16 cm.

Science Applications

Measure in SI

The metric system of measurement was developed in 1795. A modern form of the metric system, called the International System (SI), was adopted in 1960 and provides the standard measurements that all scientists around the world can understand.

The SI system is convenient because unit sizes vary by powers of 10. Prefixes are used to name units. Look at **Table 3** for some common SI prefixes and their meanings.

Table 3 Common SI Prefixes			
Prefix	**Symbol**	**Meaning**	
kilo-	k	1,000	thousand
hecto-	h	100	hundred
deka-	da	10	ten
deci-	d	0.1	tenth
centi-	c	0.01	hundredth
milli-	m	0.001	thousandth

Example How many grams equal one kilogram?

Step 1 Find the prefix *kilo* in **Table 3.**

Step 2 Using **Table 3,** determine the meaning of *kilo.* According to the table, it means 1,000. When the prefix *kilo* is added to a unit, it means that there are 1,000 of the units in a "*kilo*unit."

Step 3 Apply the prefix to the units in the question. The units in the question are grams. There are 1,000 grams in a kilogram.

Practice Problem Is a milligram larger or smaller than a gram? How many of the smaller units equal one larger unit? What fraction of the larger unit does one smaller unit represent?

Dimensional Analysis

Convert SI Units In science, quantities such as length, mass, and time sometimes are measured using different units. A process called dimensional analysis can be used to change one unit of measure to another. This process involves multiplying your starting quantity and units by one or more conversion factors. A conversion factor is a ratio equal to one and can be made from any two equal quantities with different units. If 1,000 mL equal 1 L then two ratios can be made.

$$\frac{1,000 \text{ mL}}{1 \text{ L}} = \frac{1 \text{ L}}{1,000 \text{ mL}} = 1$$

One can covert between units in the SI system by using the equivalents in **Table 3** to make conversion factors.

Example 1 How many cm are in 4 m?

Step 1 Write conversion factors for the units given. From **Table 3,** you know that 100 cm = 1 m. The conversion factors are

$$\frac{100 \text{ cm}}{1 \text{ m}} \ and \ \frac{1 \text{ m}}{100 \text{ cm}}$$

Step 2 Decide which conversion factor to use. Select the factor that has the units you are converting from (m) in the denominator and the units you are converting to (cm) in the numerator.

$$\frac{100 \text{ cm}}{1 \text{ m}}$$

Step 3 Multiply the starting quantity and units by the conversion factor. Cancel the starting units with the units in the denominator. There are 400 cm in 4 m.

$$4 \text{ m} \times \frac{100 \text{ cm}}{1 \text{ m}} = 400 \text{ cm}$$

Practice Problem How many milligrams are in one kilogram? (Hint: You will need to use two conversion factors from **Table 3.**)

Table 4 Unit System Equivalents

Type of Measurement	Equivalent
Length	1 in = 2.54 cm
	1 yd = 0.91 m
	1 mi = 1.61 km
Mass and Weight*	1 oz = 28.35 g
	1 lb = 0.45 kg
	1 ton (short) = 0.91 tonnes (metric tons)
	1 lb = 4.45 N
Volume	$1\ in^3 = 16.39\ cm^3$
	1 qt = 0.95 L
	1 gal = 3.78 L
Area	$1\ in^2 = 6.45\ cm^2$
	$1\ yd^2 = 0.83\ m^2$
	$1\ mi^2 = 2.59\ km^2$
	1 acre = 0.40 hectares
Temperature	$°C = \dfrac{(°F - 32)}{1.8}$
	$K = °C + 273$

*Weight is measured in standard Earth gravity.

Convert Between Unit Systems Table 4 gives a list of equivalents that can be used to convert between English and SI units.

Example If a meterstick has a length of 100 cm, how long is the meterstick in inches?

Step 1 Write the conversion factors for the units given. From **Table 4,** 1 in = 2.54 cm.

$$\frac{1\ in}{2.54\ cm}\ and\ \frac{2.54\ cm}{1\ in}$$

Step 2 Determine which conversion factor to use. You are converting from cm to in. Use the conversion factor with cm on the bottom.

$$\frac{1\ in}{2.54\ cm}$$

Step 3 Multiply the starting quantity and units by the conversion factor. Cancel the starting units with the units in the denominator. Round your answer based on the number of significant figures in the conversion factor.

$$100\ cm \times \frac{1\ in}{2.54\ cm} = 39.37\ in$$

The meterstick is 39.4 in long.

Practice Problem A book has a mass of 5 lbs. What is the mass of the book in kg?

Practice Problem Use the equivalent for in and cm (1 in = 2.54 cm) to show how $1\ in^3 = 16.39\ cm^3$.

Math Skill Handbook

Precision and Significant Digits

When you make a measurement, the value you record depends on the precision of the measuring instrument. This precision is represented by the number of significant digits recorded in the measurement. When counting the number of significant digits, all digits are counted except zeros at the end of a number with no decimal point such as 2,050, and zeros at the beginning of a decimal such as 0.03020. When adding or subtracting numbers with different precision, round the answer to the smallest number of decimal places of any number in the sum or difference. When multiplying or dividing, the answer is rounded to the smallest number of significant digits of any number being multiplied or divided.

Example The lengths 5.28 and 5.2 are measured in meters. Find the sum of these lengths and record your answer using the correct number of significant digits.

Step 1 Find the sum.

5.28 m	2 digits after the decimal
+ 5.2 m	1 digit after the decimal
10.48 m	

Step 2 Round to one digit after the decimal because the least number of digits after the decimal of the numbers being added is 1.

The sum is 10.5 m.

Practice Problem How many significant digits are in the measurement 7,071,301 m? How many significant digits are in the measurement 0.003010 g?

Practice Problem Multiply 5.28 and 5.2 using the rule for multiplying and dividing. Record the answer using the correct number of significant digits.

Scientific Notation

Many times numbers used in science are very small or very large. Because these numbers are difficult to work with scientists use scientific notation. To write numbers in scientific notation, move the decimal point until only one non-zero digit remains on the left. Then count the number of places you moved the decimal point and use that number as a power of ten. For example, the average distance from the Sun to Mars is 227,800,000,000 m. In scientific notation, this distance is 2.278×10^{11} m. Because you moved the decimal point to the left, the number is a positive power of ten.

The mass of an electron is about 0.000 000 000 000 000 000 000 000 000 000 911 kg. Expressed in scientific notation, this mass is 9.11×10^{-31} kg. Because the decimal point was moved to the right, the number is a negative power of ten.

Example Earth is 149,600,000 km from the Sun. Express this in scientific notation.

Step 1 Move the decimal point until one non-zero digit remains on the left.
1.496 000 00

Step 2 Count the number of decimal places you have moved. In this case, eight.

Step 3 Show that number as a power of ten, 10^8.

The Earth is 1.496×10^8 km from the Sun.

Practice Problem How many significant digits are in 149,600,000 km? How many significant digits are in 1.496×10^8 km?

Practice Problem Parts used in a high performance car must be measured to 7×10^{-6} m. Express this number as a decimal.

Practice Problem A CD is spinning at 539 revolutions per minute. Express this number in scientific notation.

Make and Use Graphs

Data in tables can be displayed in a graph—a visual representation of data. Common graph types include line graphs, bar graphs, and circle graphs.

Line Graph A line graph shows a relationship between two variables that change continuously. The independent variable is changed and is plotted on the *x*-axis. The dependent variable is observed, and is plotted on the *y*-axis.

Example Draw a line graph of the data below from a cyclist in a long-distance race.

Table 5 Bicycle Race Data	
Time (h)	**Distance (km)**
0	0
1	8
2	16
3	24
4	32
5	40

Step 1 Determine the *x*-axis and *y*-axis variables. Time varies independently of distance and is plotted on the *x*-axis. Distance is dependent on time and is plotted on the *y*-axis.

Step 2 Determine the scale of each axis. The *x*-axis data ranges from 0 to 5. The *y*-axis data ranges from 0 to 40.

Step 3 Using graph paper, draw and label the axes. Include units in the labels.

Step 4 Draw a point at the intersection of the time value on the *x*-axis and corresponding distance value on the *y*-axis. Connect the points and label the graph with a title, as shown in **Figure 20.**

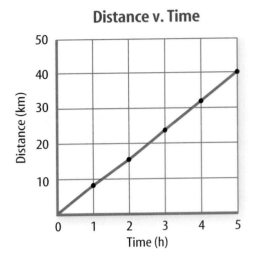

Distance v. Time

Figure 20 This line graph shows the relationship between distance and time during a bicycle ride.

Practice Problem A puppy's shoulder height is measured during the first year of her life. The following measurements were collected: (3 mo, 52 cm), (6 mo, 72 cm), (9 mo, 83 cm), (12 mo, 86 cm). Graph this data.

Find a Slope The slope of a straight line is the ratio of the vertical change, rise, to the horizontal change, run.

$$\text{Slope} = \frac{\text{vertical change (rise)}}{\text{horizontal change (run)}} = \frac{\text{change in } y}{\text{change in } x}$$

Example Find the slope of the graph in **Figure 20.**

Step 1 You know that the slope is the change in *y* divided by the change in *x*.
$$\text{Slope} = \frac{\text{change in } y}{\text{change in } x}$$

Step 2 Determine the data points you will be using. For a straight line, choose the two sets of points that are the farthest apart.
$$\text{Slope} = \frac{(40-0) \text{ km}}{(5-0) \text{ hr}}$$

Step 3 Find the change in *y* and *x*.
$$\text{Slope} = \frac{40 \text{ km}}{5 \text{ h}}$$

Step 4 Divide the change in *y* by the change in *x*.
$$\text{Slope} = \frac{8 \text{ km}}{\text{h}}$$

The slope of the graph is 8 km/h.

Bar Graph To compare data that does not change continuously you might choose a bar graph. A bar graph uses bars to show the relationships between variables. The *x*-axis variable is divided into parts. The parts can be numbers such as years, or a category such as a type of animal. The *y*-axis is a number and increases continuously along the axis.

Example A recycling center collects 4.0 kg of aluminum on Monday, 1.0 kg on Wednesday, and 2.0 kg on Friday. Create a bar graph of this data.

Step 1 Select the *x*-axis and *y*-axis variables. The measured numbers (the masses of aluminum) should be placed on the *y*-axis. The variable divided into parts (collection days) is placed on the *x*-axis.

Step 2 Create a graph grid like you would for a line graph. Include labels and units.

Step 3 For each measured number, draw a vertical bar above the *x*-axis value up to the *y*-axis value. For the first data point, draw a vertical bar above Monday up to 4.0 kg.

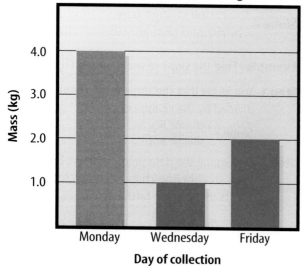

Aluminum Collected During Week

Practice Problem Draw a bar graph of the gases in air: 78% nitrogen, 21% oxygen, 1% other gases.

Circle Graph To display data as parts of a whole, you might use a circle graph. A circle graph is a circle divided into sections that represent the relative size of each piece of data. The entire circle represents 100%, half represents 50%, and so on.

Example Air is made up of 78% nitrogen, 21% oxygen, and 1% other gases. Display the composition of air in a circle graph.

Step 1 Multiply each percent by 360° and divide by 100 to find the angle of each section in the circle.

$$78\% \times \frac{360°}{100} = 280.8°$$

$$21\% \times \frac{360°}{100} = 75.6°$$

$$1\% \times \frac{360°}{100} = 3.6°$$

Step 2 Use a compass to draw a circle and to mark the center of the circle. Draw a straight line from the center to the edge of the circle.

Step 3 Use a protractor and the angles you calculated to divide the circle into parts. Place the center of the protractor over the center of the circle and line the base of the protractor over the straight line.

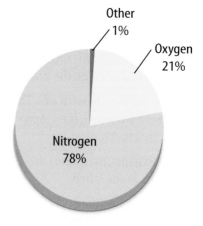

Practice Problem Draw a circle graph to represent the amount of aluminum collected during the week shown in the bar graph to the left.

Physical Science Reference Tables

Standard Units

Symbol	Name	Quantity
m	meter	length
kg	kilogram	mass
Pa	pascal	pressure
K	kelvin	temperature
mol	mole	amount of a substance
J	joule	energy, work, quantity of heat
s	second	time
C	coulomb	electric charge
V	volt	electric potential
A	ampere	electric current
Ω	ohm	resistance

Physical Constants and Conversion Factors

Acceleration due to gravity	g	9.8 m/s/s or m/s^2
Avogadro's Number	N_A	6.02×10^{23} particles per mole
Electron charge	e	1.6×10^{-19} C
Electron rest mass	m_e	9.11×10^{-31} kg
Gravitation constant	G	6.67×10^{-11} N $\times$ m^2/kg^2
Mass-energy relationship		1 u (amu) $= 9.3 \times 10^2$ MeV
Speed of light in a vacuum	c	3.00×108 m/s
Speed of sound at STP		331 m/s
Standard Pressure		1 atmosphere
		101.3 kPa
		760 Torr or mmHg
		14.7 lb/in.^{2}

Wavelengths of Light in a Vacuum

Violet	$4.0 - 4.2 \times 10^{-7}$ m
Blue	$4.2 - 4.9 \times 10^{-7}$ m
Green	$4.9 - 5.7 \times 10^{-7}$ m
Yellow	$5.7 - 5.9 \times 10^{-7}$ m
Orange	$5.9 - 6.5 \times 10^{-7}$ m
Red	$6.5 - 7.0 \times 10^{-7}$ m

The Index of Refraction for Common Substances
($\lambda = 5.9 \times 10^{-7}$ m)

Air	1.00
Alcohol	1.36
Canada Balsam	1.53
Corn Oil	1.47
Diamond	2.42
Glass, Crown	1.52
Glass, Flint	1.61
Glycerol	1.47
Lucite	1.50
Quartz, Fused	1.46
Water	1.33

Heat Constants

	Specific Heat (average) (kJ/kg $\times$ °C) (J/g $\times$ °C)	Melting Point (°C)	Boiling Point (°C)	Heat of Fusion (kJ/kg) (J/g)	Heat of Vaporization (kJ/kg) (J/g)
Alcohol (ethyl)	2.43 (liq.)	-117	79	109	855
Aluminum	0.90 (sol.)	660	2467	396	10500
Ammonia	4.71 (liq.)	-78	-33	332	1370
Copper	0.39 (sol.)	1083	2567	205	4790
Iron	0.45 (sol.)	1535	2750	267	6290
Lead	0.13 (sol.)	328	1740	25	866
Mercury	0.14 (liq.)	-39	357	11	295
Platinum	0.13 (sol.)	1772	3827	101	229
Silver	0.24 (sol.)	962	2212	105	2370
Tungsten	0.13 (sol.)	3410	5660	192	4350
Water (solid)	2.05 (sol.)	0	–	334	–
Water (liquid)	4.18 (liq.)	–	100	–	–
Water (vapor)	2.01 (gas)	–	–	–	2260
Zinc	0.39 (sol.)	420	907	113	1770

PERIODIC TABLE OF THE ELEMENTS

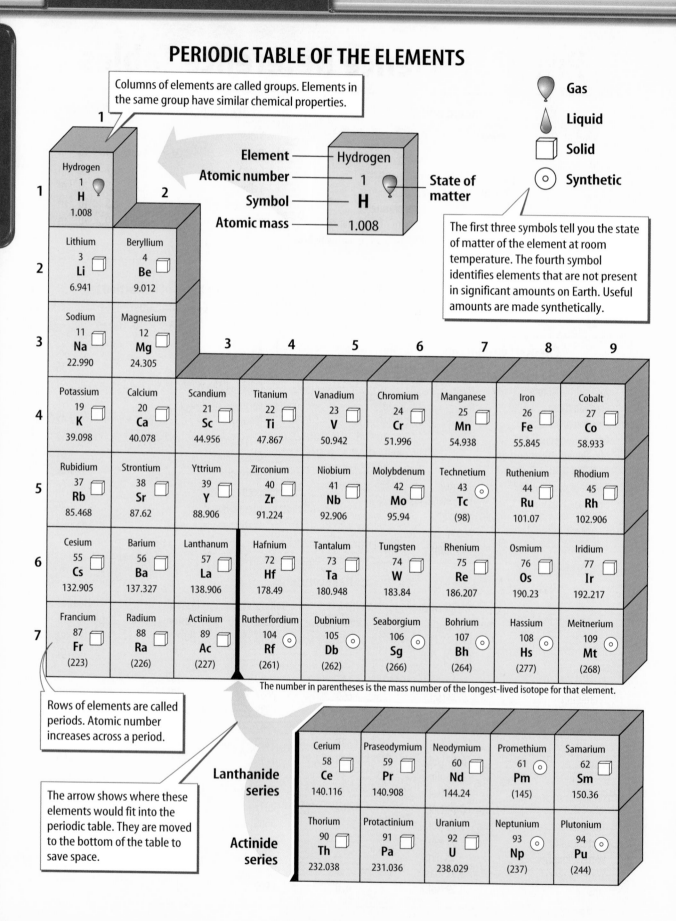

Columns of elements are called groups. Elements in the same group have similar chemical properties.

Element — Hydrogen
Atomic number — 1
Symbol — H
Atomic mass — 1.008

State of matter

Gas
Liquid
Solid
Synthetic

The first three symbols tell you the state of matter of the element at room temperature. The fourth symbol identifies elements that are not present in significant amounts on Earth. Useful amounts are made synthetically.

1									
Hydrogen 1 H 1.008	2								
Lithium 3 Li 6.941	Beryllium 4 Be 9.012								
Sodium 11 Na 22.990	Magnesium 12 Mg 24.305	3	4	5	6	7	8	9	
Potassium 19 K 39.098	Calcium 20 Ca 40.078	Scandium 21 Sc 44.956	Titanium 22 Ti 47.867	Vanadium 23 V 50.942	Chromium 24 Cr 51.996	Manganese 25 Mn 54.938	Iron 26 Fe 55.845	Cobalt 27 Co 58.933	
Rubidium 37 Rb 85.468	Strontium 38 Sr 87.62	Yttrium 39 Y 88.906	Zirconium 40 Zr 91.224	Niobium 41 Nb 92.906	Molybdenum 42 Mo 95.94	Technetium 43 Tc (98)	Ruthenium 44 Ru 101.07	Rhodium 45 Rh 102.906	
Cesium 55 Cs 132.905	Barium 56 Ba 137.327	Lanthanum 57 La 138.906	Hafnium 72 Hf 178.49	Tantalum 73 Ta 180.948	Tungsten 74 W 183.84	Rhenium 75 Re 186.207	Osmium 76 Os 190.23	Iridium 77 Ir 192.217	
Francium 87 Fr (223)	Radium 88 Ra (226)	Actinium 89 Ac (227)	Rutherfordium 104 Rf (261)	Dubnium 105 Db (262)	Seaborgium 106 Sg (266)	Bohrium 107 Bh (264)	Hassium 108 Hs (277)	Meitnerium 109 Mt (268)	

The number in parentheses is the mass number of the longest-lived isotope for that element.

Rows of elements are called periods. Atomic number increases across a period.

The arrow shows where these elements would fit into the periodic table. They are moved to the bottom of the table to save space.

	Cerium 58 Ce 140.116	Praseodymium 59 Pr 140.908	Neodymium 60 Nd 144.24	Promethium 61 Pm (145)	Samarium 62 Sm 150.36
Lanthanide series					
Actinide series	Thorium 90 Th 232.038	Protactinium 91 Pa 231.036	Uranium 92 U 238.029	Neptunium 93 Np (237)	Plutonium 94 Pu (244)

Metal

Metalloid

Nonmetal

The color of an element's block tells you if the element is a metal, nonmetal, or metalloid.

Science Online

Visit booko.msscience.com for the updates to the periodic table.

	18
	Helium 2 **He** 4.003

13	14	15	16	17	
Boron 5 **B** 10.811	Carbon 6 **C** 12.011	Nitrogen 7 **N** 14.007	Oxygen 8 **O** 15.999	Fluorine 9 **F** 18.998	Neon 10 **Ne** 20.180
Aluminum 13 **Al** 26.982	Silicon 14 **Si** 28.086	Phosphorus 15 **P** 30.974	Sulfur 16 **S** 32.065	Chlorine 17 **Cl** 35.453	Argon 18 **Ar** 39.948

10	11	12						
Nickel 28 **Ni** 58.693	Copper 29 **Cu** 63.546	Zinc 30 **Zn** 65.409	Gallium 31 **Ga** 69.723	Germanium 32 **Ge** 72.64	Arsenic 33 **As** 74.922	Selenium 34 **Se** 78.96	Bromine 35 **Br** 79.904	Krypton 36 **Kr** 83.798
Palladium 46 **Pd** 106.42	Silver 47 **Ag** 107.868	Cadmium 48 **Cd** 112.411	Indium 49 **In** 114.818	Tin 50 **Sn** 118.710	Antimony 51 **Sb** 121.760	Tellurium 52 **Te** 127.60	Iodine 53 **I** 126.904	Xenon 54 **Xe** 131.293
Platinum 78 **Pt** 195.078	Gold 79 **Au** 196.967	Mercury 80 **Hg** 200.59	Thallium 81 **Tl** 204.383	Lead 82 **Pb** 207.2	Bismuth 83 **Bi** 208.980	Polonium 84 **Po** (209)	Astatine 85 **At** (210)	Radon 86 **Rn** (222)
Darmstadtium 110 **Ds** (281)	* Unununium 111 **Uuu** (272)	* Ununbium 112 **Uub** (285)		* Ununquadium 114 **Uuq** (289)		** 116		** 118

* The names and symbols for elements 111–114 are temporary. Final names will be selected when the elements' discoveries are verified.

** Elements 116 and 118 were thought to have been created. The claim was retracted because the experimental results could not be repeated.

Europium 63 **Eu** 151.964	Gadolinium 64 **Gd** 157.25	Terbium 65 **Tb** 158.925	Dysprosium 66 **Dy** 162.500	Holmium 67 **Ho** 164.930	Erbium 68 **Er** 167.259	Thulium 69 **Tm** 168.934	Ytterbium 70 **Yb** 173.04	Lutetium 71 **Lu** 174.967
Americium 95 **Am** (243)	Curium 96 **Cm** (247)	Berkelium 97 **Bk** (247)	Californium 98 **Cf** (251)	Einsteinium 99 **Es** (252)	Fermium 100 **Fm** (257)	Mendelevium 101 **Md** (258)	Nobelium 102 **No** (259)	Lawrencium 103 **Lr** (262)

Reference Handbooks

Standard Units

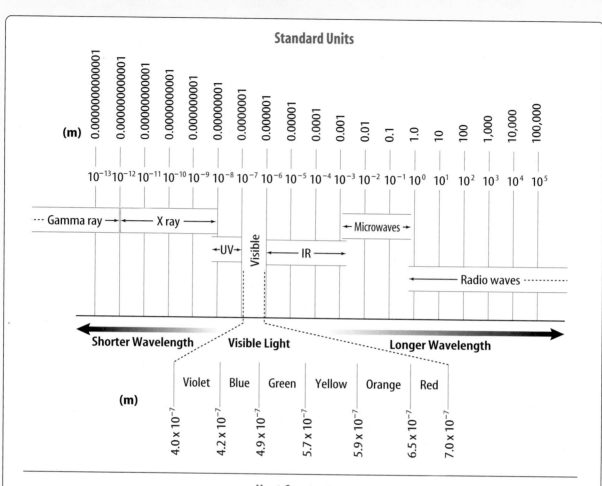

Heat Constants

Atomic number and chemical symbol

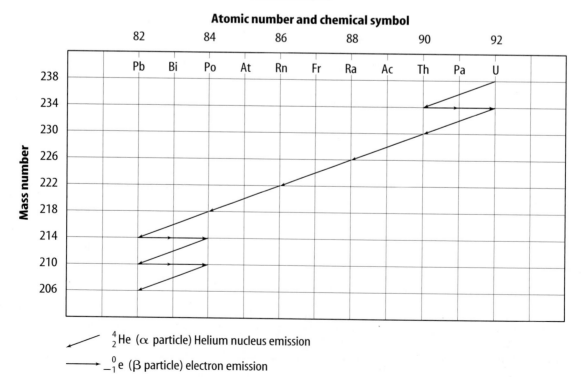

$^{4}_{2}$He (α particle) Helium nucleus emission

$^{0}_{-1}$e (β particle) electron emission

Cómo usar el glosario en español:
1. Busca el término en inglés que desees encontrar.
2. El término en español, junto con la definición, se encuentran en la columna de la derecha.

Pronunciation Key

Use the following key to help you sound out words in the glossary.

a..............back (BAK)	ew.............food (FEWD)
ay.............day (DAY)	yoo...........pure (PYOOR)
ah.............father (FAH thur)	yew...........few (FYEW)
ow.............flower (FLOW ur)	uh.............comma (CAH muh)
ar.............car (CAR)	u (+ con).......rub (RUB)
e.............less (LES)	sh.............shelf (SHELF)
ee.............leaf (LEEF)	ch.............nature (NAY chur)
ih.............trip (TRIHP)	g.............gift (GIHFT)
i (i + con + e)..idea (i DEE uh)	j.............gem (JEM)
oh.............go (GOH)	ing.............sing (SING)
aw.............soft (SAWFT)	zh.............vision (VIH zhun)
or.............orbit (OR buht)	k.............cake (KAYK)
oy.............coin (COYN)	s.............seed, cent (SEED, SENT)
oo.............foot (FOOT)	z.............zone, raise (ZOHN, RAYZ)

English — A — Español

amplitude: for a transverse wave, one half the distance between a crest and a trough. (p. 13)

amplitud: la mitad de la distancia entre la cresta y el valle en una onda transversal. (p. 13)

C

carrier wave: radio waves broadcast by a radio or TV station at an assigned frequency that contains information used to produce pictures and sound. (p. 82)

ondas conductoras: ondas de radio emitidas por una estación de radio o televisión a una frecuencia asignada, las cuales contienen información utilizada para producir imágenes y sonido. (p. 82)

compressional wave: a type of mechanical wave in which matter in the medium moves forward and backward along the direction the wave travels. (p. 11)

onda de compresión: tipo de onda mecánica en la que la materia en el medio se mueve hacia adelante y hacia atrás en dirección de la onda. (p. 11)

concave lens: lens that is thicker at its edges than in the middle. (p. 111)

lente cóncavo: lente que es más grueso en sus bordes que en el centro. (p. 111)

convex lens: lens that is thicker in the middle than at its edges. (p. 110)

ente convexo: lente que es más grueso en el centro que en sus bordes. (p. 110)

D

diffraction: bending of waves around a barrier. (p. 21)

difracción: curvatura de las ondas alrededor de una barrera. (p. 21)

Doppler effect: change in the frequency of a sound wave that occurs when the sound source and the listener are in motion relative to each other. (p. 42)

efecto Doppler: cambio en la frecuencia de una onda sonora que ocurre cuando la fuente de sonido y quien lo escucha están en movimiento relativo el uno del otro. (p. 42)

Glossary/Glosario

E

eardrum: membrane stretching across the ear canal that vibrates when sound waves reach the middle ear. (p. 54)

echo: a reflected sound wave. (p. 41)

electromagnetic spectrum: range of electromagnetic waves, including radio waves, visible light, and X rays, with different frequencies and wavelengths. (p. 71)

electromagnetic waves: waves that can travel through matter or empty space; include radio waves, infrared waves, visible light waves, ultraviolet waves, X rays and gamma rays, and are produced by moving charged particles. (pp. 12, 66)

tímpano: membrana que se extiende a través del canal auditivo y que vibra cuando las ondas sonoras alcanzan el oído medio. (p. 54)

eco: el reflejo de una onda sonora. (p. 41)

espectro electromagnético: rango de ondas electromagnéticas, incluyendo las ondas de radio, luz visible, y rayos X, con diferentes frecuencias y longitudes de onda. (p. 71)

ondas electromagnéticas: ondas que pueden viajar a través de la materia o del espacio vacío; incluyen ondas radiales, ondas infrarrojas, ondas de luz visible, ondas ultravioletas, rayos X y rayos gama, y que son producidas por partículas cargadas en movimiento. (pp. 12, 66)

F

focal length: distance along the optical axis from the center of a mirror or lens to the focal point. (p. 104)

focal point: point on the optical axis of a mirror or lens where rays traveling parallel to the optical axis pass through. (p. 104)

frequency: number of wavelengths that pass a given point in one second; measured in hertz (Hz). (p. 15)

fundamental frequency: lowest natural frequency that is produced by a vibrating object, such as a string or a column of air. (p. 49)

distancia focal: distancia a lo largo del eje óptico desde el centro de un espejo o lente hasta el punto focal. (p. 104)

punto focal: punto en el eje óptico de un espejo o lente por el cual atraviesan los rayos que viajan en paralelo al eje óptico. (p. 104)

frecuencia: número de longitudes de onda que pasan un punto determinado en un segundo; se mide en hertz (Hz). (p. 15)

recuencia fundamental: frecuencia natural más baja producida por un objeto que vibra, tal como una cuerda o una columna de aire. (p. 49)

G

gamma ray: highest-energy electromagnetic waves with the shortest wavelengths and highest frequencies. (p. 76)

Global Positioning System (GPS): uses satellites, ground-based stations, and portable units with receivers to locate objects on Earth. (p. 85)

rayos gama: ondas electromagnéticas que poseen la mayor cantidad de energía y las cuales presentan las longitudes de onda más cortas y las frecuencias más altas. (p. 76)

Sistema de Posicionamiento Global (SPG): sistema que utiliza satélites, estaciones en tierra y unidades portátiles con receptores para ubicar objetos en la Tierra. (p. 85)

I

infrared wave: electromagnetic waves with wavelengths between 1 mm and 0.7 millionths of a meter. (p. 73)

ondas infrarrojas: ondas electromagnéticas con longitudes de onda entre un milímetro y 0.7 millonésimas de metro. (p. 73)

Glossary/Glosario

interference: occurs when two or more waves combine and form a new wave when they overlap. (p. 23)

interferencia: ocurre cuando dos o más ondas se combinan y al sobreponerse forman una nueva onda. (p. 23)

law of reflection: states that when a wave is reflected, the angle of incidence is equal to the angle of reflection. (p. 101)

lens: transparent object that has at least one curved surface that causes light to bend. (p. 109)

light ray: narrow beam of light traveling in a straight line. (p. 96)

loudness: the human perception of how much energy a sound wave carries. (p. 38)

ley de la reflexión: establece que cuando se refleja una onda, el ángulo de incidencia es igual al ángulo de reflexión. (p. 101)

lente: objeto transparente que tiene por lo menos una superficie curva que hace cambiar la dirección de la luz. (p. 109)

rayo de luz: haz estrecho de luz que viaja en línea recta. (p. 96)

intensidad: percepción humana de la cantidad de energía conducida por una onda sonora. (p. 38)

mechanical wave: a type of wave that can travel only through matter. (p. 9)

medium: material through which a wave travels. (p. 97)

onda mecánica: tipo de onda que puede viajar únicamente a través de la materia. (p. 9)

medio: material a través del cual viaja una onda. (p. 97)

natural frequencies: frequencies at which an object will vibrate when it is struck or disturbed. (p. 47)

frecuencias naturales: frecuencias a las cuales un objeto vibrará cuando es golpeado o perturbado. (p. 47)

overtones: multiples of the fundamental frequency. (p. 49)

armónicos: múltiplos de la frecuencia fundamental. (p. 49)

pitch: how high or low a sound is. (p. 40)

altura: expresa qué tan alto o bajo es un sonido. (p. 40)

radiant energy: energy carried by an electromagnetic wave. (p. 70)

energía radiante: energía conducida por una onda electromagnética. (p. 70)

radio waves: lowest-frequency electromagnetic waves that have wavelengths greater than about 0.3 m and are used in most forms of telecommunications technology—such as TVs, telephones, and radios. (p. 72)

reflecting telescope: uses a concave mirror to gather light from distant objects. (p. 115)

reflection: occurs when a wave strikes an object or surface and bounces off. (p. 19)

refracting telescope: uses two convex lenses to gather light and form an image of a distant object. (p. 114)

refraction: bending of a wave as it moves from one medium into another medium. (p. 20)

resonance: occurs when an object is made to vibrate at its natural frequencies by absorbing energy from a sound wave or other object vibrating at this frequency. (p. 48)

reverberation: repeated echoes of sounds. (p. 53)

ondas de radio: ondas electromagnéticas con la menor frecuencia, las cuales poseen longitudes de onda mayores de unos 0.3 metros y son utilizadas en la mayoría de técnicas de telecomunicaciones, tales como televisores, teléfonos y radios. (p. 72)

telescopio de reflexión: utiliza un espejo cóncavo para concentrar la luz proveniente de objetos lejanos. (p. 115)

reflexión: ocurre cuando una onda choca contra un objeto o superficie y rebota. (p. 19)

telescopio de refracción: utiliza dos lentes convexos para concentrar la luz y formar una imagen de un objeto lejano. (p. 114)

refracción: curvatura de una onda a medida que se mueve de un medio a otro. (p. 20)

esonancia: ocurre cuando se hace vibrar un objeto a sus frecuencias naturales mediante la absorción de energía de una onda sonora o de otro objeto que vibra a dicha frecuencia. (p. 48)

reverberación: ecos repetidos de los sonidos. (p. 53)

transverse wave: a type of mechanical wave in which the wave energy causes matter in the medium to move up and down or back and forth at right angles to the direction the wave travels. (p. 10)

onda transversal: tipo de onda mecánica en el cual la energía de la onda hace que la materia en el medio se mueva hacia arriba y hacia abajo o hacia adelante y hacia atrás en ángulos rectos respecto a la dirección en que viaja la onda. (p. 10)

ultraviolet radiation: electromagnetic waves with wavelengths between about 0.4 millionths of a meter and 10 billionths of a meter; has frequencies and wavelengths between visible light and X rays. (p. 75)

radiación ultravioleta: ondas electromagnéticas con longitudes de onda entre aproximadamente 0.4 millonésimas de metro y 10 billonésimas de metro; tienen frecuencias y longitudes de onda entre aquellas de la luz visible y los rayos X. (p. 75)

visible light: electromagnetic waves with wavelengths between 0.4 and 0.7 millionths of a meter that can be seen with your eyes. (p. 74)

luz visible: ondas electromagnéticas con longitudes de onda entre 0.4 y 0.7 millonésimas de metro y que pueden ser observadas a simple vista. (p. 74)

W

wave: rhythmic disturbance that carries energy but not matter. (p. 8)

wavelength: for a transverse wave, the distance between the tops of two adjacent crests or the bottoms of two adjacent troughs; for a compressional wave, the distance from the centers of adjacent rarefactions or adjacent compressions. (p. 14)

onda: alteración rítmica que transporta energía pero no materia. (p. 8)

longitud de onda: en una onda transversal, es la distancia entre las puntas de dos crestas adyacentes o entre dos depresiones adyacentes; en una onda de compresión es la distancia entre los centros de dos rarefacciones adyacentes o compresiones adyacentes. (p. 14)

X

X ray: high-energy electromagnetic wave that is highly penetrating and can be used for medical diagnosis. (p. 76)

rayos X: ondas electromagnéticas de alta energía, las cuales son altamente penetrantes y pueden ser utilizadas para diagnósticos médicos. (p. 76)

Index

Credits

Magnification Key: Magnifications listed are the magnifications at which images were originally photographed.
LM–Light Microscope
SEM–Scanning Electron Microscope
TEM–Transmission Electron Microscope

Acknowledgments: Glencoe would like to acknowledge the artists and agencies who participated in illustrating this program: Absolute Science Illustration; Andrew Evansen; Argosy; Articulate Graphics; Craig Attebery, represented by Frank & Jeff Lavaty; CHK America; John Edwards and Associates; Gagliano Graphics; Pedro Julio Gonzalez, represented by Melissa Turk & The Artist Network; Robert Hynes, represented by Mendola Ltd.; Morgan Cain & Associates; JTH Illustration; Laurie O'Keefe; Matthew Pippin, represented by Beranbaum Artist's Representative; Precision Graphics; Publisher's Art; Rolin Graphics, Inc.; Wendy Smith, represented by Melissa Turk & The Artist Network; Kevin Torline, represented by Berendsen and Associates, Inc.; WILDlife ART; Phil Wilson, represented by Cliff Knecht Artist Representative; Zoo Botanica.

Photo Credits

Cover V. Cary Wolinsky/Stock Boston/PictureQuest; **i ii** V. Cary Wolinsky/Stock Boston/PictureQuest; **iv** (bkgd)John Evans, (inset)V. Cary Wolinsky/Stock Boston/ PictureQuest; **v** (t)PhotoDisc, (b)John Evans; **vi** (l)John Evans, (r)Geoff Butler; **vii** (l)John Evans, (r)PhotoDisc; **viii** PhotoDisc; **ix** Aaron Haupt Photography; **x** (t)Ken Frick, (b)Matt Meadows; **xi** Susumu Nishinaga/Science Photo Library/Photo Researchers, Inc.; **xii** (t)David Young-Wolff/ PhotoEdit, (b)Steven Starr/Stock Boston; **1** Roger Ressmeyer/ CORBIS; **2** Bettmann/CORBIS; **3** (t)Schnectady Museum; Hall of Electrical History Foundation/CORBIS, (bl)U.S. Department of the Interior, National Park Service, Edison National Historic Site, (br)CORBIS; **4** Schnectady Museum; Hall of Electrical History Foundation/CORBIS; **6–7** Douglas Peebles/CORBIS; **8** (l)file photo, (r)David Young-Wolff/ PhotoEdit, Inc.; **9** David Young-Wolff/PhotoEdit, Inc.; **10** Mark Thayer; **13** Steven Starr/Stock Boston; **18** Ken Frick; **19** Mark Burnett; **21** Ernst Haas/Stone/Getty Images; **22** Peter Beattie/Liaison Agency/Getty Images; **24** D. Boone/CORBIS; **25** Seth Resnick/Stock Boston; **26–27** John Evans; **28** (t)Roger Ressmeyer/CORBIS, (b)SuperStock; **32** Mark Burnett; **34–35** Tom Wagner/CORBIS SABA; **39** (t)Joe Towers/The Stock Market/CORBIS, (c)Bob Daemmrich/Stock Boston/ PictureQuest, (b)Jean-Paul Thomas/Jacana Scientific Control/Photo Researchers; **42** NOAA; **45** Spencer Grant/ PhotoEdit, Inc.; **46** Timothy Fuller; **50** Dilip Mehta/Contact Press Images/PictureQuest; **51** (tr)Paul Seheult/Eye Ubiquitous/CORBIS, (b)Icon Images, (tl)CORBIS; **52** (t)William Whitehurst/The Stock Market/CORBIS, (b)G. Salter/Lebrecht Music Collection; **53** SuperStock; **54** (t)Geostock/PhotoDisc, (b)SuperStock; **55** Fred E. Hossler/Visuals Unlimited; **56** (t)Will McIntyre/Photo Researchers, (b)Oliver Benn/Stone/Getty Images; **58** Douglas Whyte/The Stock Market/CORBIS; **59** (l)The Photo Works/ Photo Researchers, (r)PhotoDisc; **61** C. Squared Studios/ PhotoDisc; **64–65** Maxine Hall/CORBIS; **66** (l)Bob Abraham/ The Stock Market/CORBIS, (r)Jeff Greenberg/Visuals Unlimited; **67** (l)David Young-Wolff/PhotoEdit, Inc., (r)NRSC, Ltd./Science Photo Library/Photo Researchers; **68** (t)Grantpix/Photo Researchers, (b)Richard Megna/ Fundamental Photographs; **70** Luke Dodd/Science Photo Library/Photo Researchers; **72** (t)Matt Meadows, (b)Jean Miele/The Stock Market/CORBIS; **74** (t)Gregory G. Dimijian/Photo Researchers, (b)Charlie Westerman/Liaison/ Getty Images; **75** Aaron Haupt; **77** (l)Matt Meadows, (r)Bob Daemmrich/The Image Works; **78** (tr)Phil Degginger/Color-Pic, (l)Phil Degginger/NASA/Color-Pic, (cr)Max Planck Institute for Radio Astronomy/Science Photo Library/Photo Researchers, (br)European Space Agency/Science Photo Library/Photo Researchers; **79** (l)Harvard-Smithsonian Center for Astrophysics, (c)NASA/Science Photo Library/ Photo Researchers, (r)European Space Agency; **80** Timothy Fuller; **85** Ken M. Johns/Photo Researchers; **86** (t)Michael Thomas/Stock South/PictureQuest, (b)Dominic Oldershaw; **87** Michael Thomas/Stock South/PictureQuest; **88** (bkgd)TIME, (t)Culver Pictures, (b)Hulton Archive/Getty Images; **89** (l)Macduff Everton/CORBIS, (r)NASA/Mark Marten/Photo Researchers; **93** Eric Kamp/Index Stock; **94–95** Chad Ehlers/Index Stock; **96** Dick Thomas/Visuals Unlimited; **97** John Evans; **98** (tl)Bob Woodward/The Stock Market/CORBIS, (tc)Ping Amranand/Pictor, (tr)SuperStock, (b)Runk/Schoenberger from Grant Heilman; **99** Mark Thayer; **102** (l)Susumu Nishinaga/Science Photo Library/ Photo Researchers, (r)Matt Meadows; **104** (l)Matt Meadows, (r)Paul Silverman/Fundamental Photographs; **105** (l)Digital Stock, (r)Joseph Palmieri/Pictor; **107** Geoff Butler; **109** Richard Megna/Fundamental Photographs; **113** David Young-Wolff/PhotoEdit, Inc.; **114 115** Roger Ressmeyer/ CORBIS; **118 119** Geoff Butler; **120** The Stapleton Collection/Bridgeman Art Library; **126** PhotoDisc; **128** Tom Pantages; **132** Michell D. Bridwell/PhotoEdit, Inc.; **133** (t)Mark Burnett, (b)Dominic Oldershaw; **134** StudioOhio; **135** Timothy Fuller; **136** Aaron Haupt; **138** KS Studios; **139** Matt Meadows; **141** (t)Mark Burnett, (b)Amanita Pictures; **142** Amanita Pictures; **143** Bob Daemmrich; **145** Davis Barber/PhotoEdit, Inc.

PERIODIC TABLE OF THE ELEMENTS

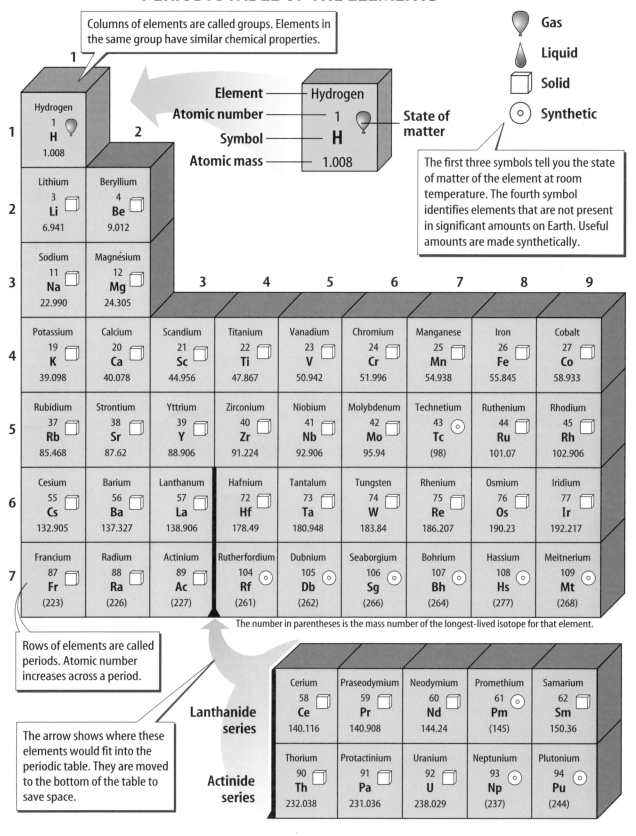